An Introduction to

Critical Path Analysis

K. G. Lockyer

Professor of Operations Management
University of Bradford

THIRD EDITION

PITMAN

Fordham University
LIBRARY
AT
LINCOLN CENTER
New York, N. Y.

PITMAN BOOKS LIMITED
128 Long Acre, London WC2E 9AN

PITMAN PUBLISHING INC
1020 Plain Street, Marshfield, Massachusetts 02050

Associated Companies
Pitman Publishing Pty Ltd, Melbourne
Pitman Publishing New Zealand Ltd, Wellington
Copp Clark Ltd, Toronto

© K. G. Lockyer 1964, 1967, 1969

Third Edition 1969, Reprinted 1974, 1975, 1978, 1980, 1982 (twice)
First published in paperback 1977

All rights reserved. No part of this publication may be reproduced,
stored in a retrieval system, or transmitted, in any form or by any
means, electronic, mechanical, photocopying, recording and/or
otherwise without the prior written permission of the publishers.
This book may not be lent, resold, hired out or otherwise disposed of
by way of trade in any form of binding or cover other than that in
which it is published, without the prior consent of the publishers.
This book is sold subject to the Standard Conditions of Sale
of Net Books and may not be resold in the UK below the net price.

Printed and bound in Great Britain
at The Pitman Press, Bath

ISBN 0 273 01177 4

T
57
.85
.L6
1977

AN INTRODUCTION TO
CRITICAL PATH ANALYSIS

PREFACE TO THIRD EDITION

THE purpose of this text is, as stated in the previous editions, essentially a practical one. It is intended both for those who wish to assess the value of Critical Path Analysis in dealing with their own problems and also for those who, being convinced of its value, wish to put the technique into practice. It is therefore written "from the ground up," assuming no prior knowledge on the part of the reader, and requiring no mathematical expertise. Experienced workers in this field may find that some of the discussion is apparently excessively long or excessively simple, but the subject matter has been shaped by a very large number of lectures and consultancy assignments undertaken by myself and my colleagues throughout the world. This work has highlighted the initial difficulties and problems encountered by students, and these have been elaborated upon as far as possible within a working book.

In a subject exploding as violently as CPA there is a temptation to include in a textbook every new idea, manipulation and treatment without regard to their practical value. Two tests have been applied to all new material—

    (i) is it useful?
    (ii) is it proven?

Failure to meet these criteria has caused new ideas, however elegant or intellectually appealing, to be rejected.

This edition differs from the previous edition in the following respects—

Dummy activities are treated at length in a special chapter (Chapter 3) which includes a "foolproof" method of locating dummies. The "ladder" method of dealing with overlapping activities in very thoroughly discussed, again in a special chapter (Chapter 7), which probably constitutes the most exhaustive examination of this method generally available. The problems of reducing the project time are amplified, and yet another method of deriving a bar chart is given. Each of the methods shown has its own particular virtues and no one method is recommended above any other and use is deliberately made of all methods.

There are three wholly new chapters: Chapter 11 gives a manual method of resource allocation and tribute must be made here to the excellent work of R. L. Martino in this and associated fields. Chapter 13 treats with some of the interesting Activity-on-Node systems of Networking. These are coming to be used both in the U.K. and on the Continent, and a number of U.S. writers have been arguing powerfully—but apparently unsuccessfully—for wider adoption of these methods in the U.S. Chapter 14 deals with Line of Balance, showing it to be most simply regarded as a CPA system.

Wherever possible the definitions appearing in B.S. 4335:1968 are used. In the cases where I have felt that the recommended definitions are in some way incorrect or inadequate, my own, as well as BS 4335:1968 definitions are given.

The subject matter being new and in a ferment, it is not surprising to find that existing practitioners are enthusiasts who are willing to discuss their experiences. I have derived much benefit from these discussions and should like to thank particularly David Armstrong (formerly of Richard Costain Ltd.), Brian Fine (formerly of Constructors John Brown Ltd.) and Fred Moon (Southern Project Group, Central Electricity Generating Board), and P. A. Rhodes (I.C.L. Ltd.) from all of whom I have learned a great deal. It may well be that they will find some of their own ideas appearing in the following pages without appropriate acknowledgement. If so, it is quite unintentional and I should like to add my apologies to my thanks. My many colleagues in Lockyer & Partners Ltd. and Harold Whitehead & Partners Ltd. have assisted me enormously by sharing with me their practical experience in using this most versatile technique.

I should also like to thank those many readers who have written pointing out errors and obscurities. Any remaining faults, of course, are entirely my own responsibility.

K. G. LOCKYER

CONTENTS

ix

CHAPTER ONE

INTRODUCTORY

WHEN attempting to determine the completion date for any task, whether it be the building of a bridge, the mounting of a sales conference, the designing of a new piece of equipment or any other project, it is necessary to time-table all the activities which make up the task, that is to say, a *plan* must be prepared. The need for planning has always been present, but the complexity and competitiveness of modern undertakings now requires that this need should be met rather than just recognized.

Characteristics of Effective Planning

Present decisions affect both present *and future* actions, and if immediate, short-term decisions are not taken within the framework of long-term plans, then the short-term decisions may effectively impose some long-term actions which are undesirable but inescapable. Military writers categorize these long- and short-term plans as—

    strategical plans, which are those made to ". . . serve the needs of generalship"; and

    tactical plans, which are those made ". . . when in contact with the enemy."

Ideally, of course strategical plans are made before the start of an operation and by following them, the operation is successfully concluded. Inevitably however, tactical decisions will have to be made and these can only be successful if they are made within the context of the strategic plan.

To permit effective tactical decision-making, it is necessary that the strategy shall be expressed in a form which is—

(1) explicit,
(2) intelligible,
(3) capable of accepting change,
(4) capable of being monitored.

It is suggested that these are the minimum acceptable characteristics of an effective plan. Given these characteristics, tactical decisions can then be taken which have results which are both *predictable* and *acceptable* to the strategic needs.

It must be realized that there is no absolute definition of strategy and tactics: at any one level in a hierarchy a tactical plan should be made to fit the needs of the strategy received from a higher level, and this tactical plan then becomes the strategy for a lower level. Freedom to make appropriate decisions must be given to those who will be held responsible for performance, and if these decisions are to be meaningful in a larger context, then it is imperative that this larger context shall be known. Freedom "within the law" is as important a concept in management as in the community. Too often, tactical decisions are taken to meet the needs only of immediate expediency, and the results may well be disastrous in the long term.

An Effective Plan Will Be
(i) explicit
(ii) intelligible
(iii) capable of accepting change
(iv) capable of being monitored

Historical Background to CPA

Before the advent of CPA, probably the best-known way of trying to plan was by means of a bar or Gantt chart, and although this is extremely useful in many cases, it suffers from an inability to show the inter-relationships between the various activities. Thus, it is not possible to deduce from a Gantt chart that activity X must be complete before activity Y can be started, or that a delay between activity Y and activity Z is permissible but not essential. In small projects this is not serious, as the planner can remember the various links between activities, but in large projects such feats of memory are impossible, and the Gantt charting technique is then of very limited value.

The middle of the 1950s saw an explosion in interest in this problem. In Great Britain the Operational Research Section of the Central Electricity Generating Board investigated the problems concerned with the overhaul of generating plant—a task of considerable complexity which was increasing in importance as new high-performance plant was being brought into service. By 1957 the O.R. Section had devised a technique which consisted essentially of identifying the "longest irreducible sequence of events," and using this technique they carried out in 1958 an experimental overhaul at a power station which reduced the overall time to 42 per cent of the previous average time for the same work. Continuing to work upon these lines the overhaul time was further reduced by 1960 to 32 per cent of the previous average time. The rather clumsy name, "longest irreducible sequence of events," was soon replaced by the name, "major sequence," and it was pointed out, for example, that delays in the "major sequence" would delay completion times, but that difficulties elsewhere need not necessarily involve extensions in total time. This work of the O.R. group was not made public, although comprehensive reports were circulated internally which foreshadowed much later work carried out elsewhere.

At much the same time work was being undertaken in the U.S.A., and in early 1958 the U.S. Navy Special Projects Office set up a team to devise a means of dealing with the planning and subsequent control of complex work. This investigation was known as the Program Evaluation Research Task, which gave rise to (or possibly derived from) the code name PERT. By February, 1958, Dr. C. E. Clark, a mathematician in the PERT team, presented the early notions of arrow-diagramming, doubtless drawing from his study of graphics. This early work of Dr. Clark was rapidly polished and by July, 1958, the first report, *PERT, Summary Report, Phase I*, was published. By this time the full title of the work had become *Program Evaluation and Review Technique*, and the value of the Technique seemed well established. By October, 1958, it was decided to apply PERT to the Fleet Ballistic Missiles Programme, where it was credited with saving two years in the development of the Polaris missile.

Similar development work was being undertaken elsewhere— for example in the U.S. Air Force under the code name PEP. Also in 1958, the E.I. du Pont de Nemours Company used a technique called the Critical Path Method (CPM) to schedule and control a very large project, and during the first complete year's use of CPM it was credited with saving the company $1 million. Subsequent use underlined the basic simplicity and extraordinary usefulness of this method, and by 1959 Dr. Mauchly, who had worked on the Du Pont project, set up an organization to solve industrial problems using the Critical Path Method.

Since 1958 considerable work has been carried out, mainly in the United States of America, in consolidating and improving these techniques. Much of the effort has been expended by the computer companies, who have devised special names to distinguish their own work, these names taking on almost the air of trade-marks. Whilst this concentration of effort is very desirable, two unfortunate results have accrued. Firstly the student is faced with a large number of names for what is basically the same technique, and indeed one writer claims to have identified over forty different names in current use. The second undesirable result is that an impression has been given that CPA is a technique which is applicable only to large projects, and that it is vital to use a computer to achieve worth-while results. This is not necessarily so: quite striking results can be achieved in small projects which can be computed "by hand."

In 1968 the British Standards Institution published a standard, BS 4335:1968 *Glossary of Terms used in Project Network*

Analysis which sets out to provide a "basic glossary of terms which are currently in common use . . . [This] glossary is a minimal one, confining itself to essential terms." It recognizes that the subject is new and expanding, noting that further terms will be added when practice shows them to be necessary. The preferred term for the subject matter of the present text is Critical Path Analysis and this, or its abbreviation (CPA), is the name which will be used in this volume. Wherever possible, the other recommendations of BS 4335:1968 will be followed.

Where CPA can be Used

Critical Path Analysis can be used in situations where the start and the finish of the task can be identified: continuous or flow production is not susceptible to planning by CPA *although the setting up of pre-production work is*. The size of the project is of no consequence—CPA has been used to plan a simple test procedure just as successfully as to plan the construction of a town or the launching of a space ship.

It is quite impossible to list all the applications of CPA, since it is now used extremely widely. However, in order to give some idea of the "spread" of CPA, the following is a brief summary of some uses of which the author or his colleagues have personal knowledge. It must be emphasized that this list is not exhaustive: new applications are continually being found.

(1) *Overhaul*. Plant, equipment, vehicles and buildings, both on a routine and an emergency basis.

(2) *Construction*. Houses, flats and offices, including all pre-contract, tendering and design work.

(3) *Civil Engineering*. Motorways, bridges, road programmes, including all pre-contract, tendering and design work.

(4) *Town Planning*. Control of tendering and design procedures and subsequent building and installation of services.

(5) *Marketing*. Market research, product launching and the setting-up and running of advertising campaigns.

(6) *Ship Building*. Design and production of ships.

(7) *Design*. Design of cars, machine tools, guided weapons, computers, electronic equipment.

(8) *Pre-production*. Control of production of jigs, tools, fixtures and test equipment.

(9) *Product Change-over*. The changing over from one product, or family of products to another, for example, "winter" to "summer" goods.

(10) *Commissioning and/or Installation*. Power generation equipment of all types, and data processing plant.

(11) *Modification Programmes*. The modification of existing plant or equipment.

(12) *Office Procedures*. Investigations into existing administrative practices (for example, the preparation of monthly accounts) and the devising and installing of new systems.

(13) *Consultancy*. The setting up and control of consultancy assignments.

In the above, CPA has been used to plan and control time, the use of resources and the expenditure of capital. Attempts are being made to control costs, and whilst some of these have been extremely successful, difficulty is often experienced in first defining the elements of cost and then subsequently extracting and recording them.

The Basic Essentials

In essence, Critical Path Analysis can be considered to proceed in three distinct phases, namely *Planning, Analysing and Scheduling*, and *Controlling*. Whilst it is convenient to consider each phase separately, it is of course true that they are not independent. Thus, after initially planning a project, the subsequent analysis may show that the original plan is unacceptable and a new one must be prepared which is again modified. This "plan—test—modify—replan" sequence is general to any type of planning, but the technique of CPA allows changes to be carried out very much more readily than with any other technique.

Planning

In the first phase, *Planning*, the project being considered is represented graphically by a diagram built up from circles (representing events) and arrows (representing activities) which lead up to, or emerge from, the circles. To consider a very simple example, suppose the project involves three persons leaving place A and subsequently meeting at place B. One person goes directly from A to B, two move from A to C, one then proceeding directly to B, the other proceeding from C to D and thence to B. This would be represented on a diagram as follows—

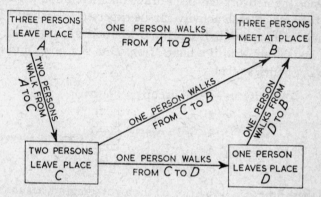

This could then be simplified to the diagram (now known as a network) shown below—

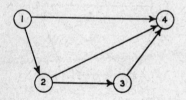

where Event 1 represents "Leave place A"

       „   2   „   "   „   „ C"

       „   3   „   "   „   „ D"

       „   4   „   "Arrive   „ B"

and Activity 1—2 represents "Walk from A to C"

       ,,   2—3   ,,   "   ,, ,, C ,, D"

       ,,   1—4   ,,   "   ,, ,, A ,, B"

       ,,   3—4   ,,   "   ,, ,, D ,, B"

       ,,   2—4   ,,   "   ,, ,, C ,, B"

It must be clearly understood that the positions of the arrows and circles bear no relation to the locations of places A, B, C and D. The arrow diagram is drawn in any manner which is geometrically convenient, and it is not (except metaphorically) a map.

Once the network has been drawn, consideration is given to the time which each activity can be expected to occupy. Thus it may be that to walk from A to C will take 2 hours, from C to D 1 hour, from A to B 2 hours, from D to B 1 hour, and from C to B 3 hours, so that

      Activity 1—2 will occupy 2 hours

           ,,   2—3   ,,   ,,   1 hour

           ,,   1—4   ,,   ,,   2 hours

           ,,   3—4   ,,   ,,   1 hour

           ,,   2—4   ,,   ,,   3 hours,

and these times can be written on the arrow diagram as subscripts on the activity arrows—

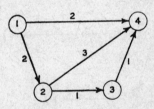

It is worth noting here that planning by CPA enables two decisions—one concerning inter-relationships, the other concerning time—to be separated, whilst other planning methods, such as the Gantt Chart, require these decisions to be made simultaneously. This separation constitutes one of the great strengths of CPA: statements of inter-relationships can be made and their logic tested without any decisions on time being made.

Analysing and Scheduling

The planning having been done, the analysis can now be carried out. First, the question "How long will the whole project take?" can be answered. Clearly, the shortest possible time which can elapse between all the men leaving place A and meeting at place B is determined by that series of activities which occupies the longest time. There are in this example three series of activities, namely

$$
\begin{aligned}
&\text{(i)} \quad 1\text{---}4 \\
&\text{(ii)} \quad 1\text{---}2\text{---}4 \\
&\text{(iii)} \quad 1\text{---}2\text{---}3\text{---}4
\end{aligned}
$$

and their durations are seen to be

$$
\begin{aligned}
&\text{(i)} \quad 2 &\text{hours} \;&= 2 \text{ hours} \\
&\text{(ii)} \quad 2 + 3 &\text{''} \;&= 5 \text{ hours} \\
&\text{(iii)} \quad 2 + 1 + 1 &\text{''} \;&= 4 \text{ hours}
\end{aligned}
$$

so that, assuming that all went according to plan, the shortest possible time which could elapse is 5 hours, and this is determined by sequence $1\text{---}2\text{---}4$. If either activity in *this* sequence were to increase, the overall project time would increase, whereas the other activities could increase substantially in time without affecting the overall time. This determining sequence is critical to the performance of the project, and is hence known as the *Critical Path*.

Isolating the Critical Path enables effort to be concentrated in the most useful way. For example, if this overall time of 5 hours is not acceptable, and it is necessary to reduce it, then "speeding up" must occur in the sequence $1\text{---}2\text{---}4$. If activity $2\text{---}3$ is reduced from 1 hour to $\frac{1}{2}$ hour, the overall time will not be affected. On the other hand, any reduction in either activity

1—2 or activity 2—4 will reduce the project time. By investigating this sequence, it is therefore possible to decide in what manner resources should be deployed to reduce project time.

If the overall time is accepted, then it is clear that activity 1—4 can increase from 2 hours to 5 hours without effect on the overall time. One way of expressing this is to say that activity 1—4 has a "float" of 3 hours; alternatively it may be said that if the whole project starts at hour 0, activity 1—4 can start anywhere between hour 0 and hour 3. If it starts as early as possible (that is, at hour 0) then it will finish as early as possible, at hour 2, whilst if it starts as *late* as possible (that is, at hour 3) then it will finish as late as possible, at hour 5. This can be summarized as follows—

Activity	Duration	Start Earliest	Start Latest	Finish Earliest	Finish Latest	Float
1—4	2	0	3	2	5	3

Similar analyses may be carried out for all activities, and a table drawn up—

Activity	Duration	Start Earliest	Start Latest	Finish Earliest	Finish Latest	Float
1—2	2	0	0	2	2	0
1—4	2	0	3	2	5	3
2—3	1	2	3	3	4	1
2—4	3	2	2	5	5	0
3—4	1	3	4	4	5	1

where it will be seen that the activities in the Critical Path possess no float.

From this analysis it is possible to assign calendar times to the various activities, and this is the process known as *scheduling*. It may be that for some reason the start of activity 1—4 is deliberately set at hour 2, and so the schedule would show the activity starting at hour 2 rather than being permitted to start between hours 0 and 3. The considerations determining the schedule are often concerned with the deployment of the various resources available.

Controlling

This analysis enables the project to be controlled while it is being undertaken. A note of the actual performance compared with the plan summarized above will show whether the overall project time is going to be achieved. For example if, after 2 hours, activity 1—2 is not complete (that is, the two persons have not arrived at place C), then the project will take longer than planned unless this "slip" is made good, and this can only be done by reducing the time in activity 2—4. Reducing activity 3—4, for example, has no useful effect at all. On the other hand if, at hour 2, activity 1—2 *is* complete, but activity 2—3 is not yet started, then it is known that this activity must be started within the next hour.

The above discussion of a very simple situation is intended to illustrate the general CPA approach. It has been possible to carry out all analyses "by inspection" ; later chapters will show how to analyse a project systematically and simply, using elementary arithmetic. Any complexity which arises will come from an increase in complexity of the network, not from the basic ideas behind this technique.

Introducing CPA into an Organization

As with any other new managerial tool, CPA will require to be introduced into an organization with care. It is suggested that the following points should be observed.

(i) CPA is not a universal tool – there are situations where it cannot be usefully employed. These situations are, in general, those where activity is continuous, for example, flow production. A CPA-type situation is characteristically one which has a definable start and a definable finish.

(ii) CPA is not a panacea – it does not cure all ills. Indeed, CPA *in itself* does not solve any problems, but it does expose situations in a way which will permit effective examinations both of the problems and of the effect of possible solutions.

However, the formulating and implementing of any solution will remain the responsibility of the appropriate manager.

(iii) CPA must not be made a mystery, known only to a chosen few. All levels require to appreciate the method and its limitations, and an extensive educational programme will be necessary to ensure that knowledge is spread as widely as possible.

(iv) The person initiating CPA into an organization must be of sufficient stature and maturity to be able to influence both senior and junior personnel.

(v) Wherever possible, the early applications of CPA should be to simple situations. If CPA is first employed on a very difficult task it may fail, not because of the difficulty of CPA but because of the difficulty of the task itself. However, the failure is likely to be attributed to CPA and the technique will be discredited.

(vi) The first application of CPA will undoubtedly excite considerable interest, and attract many resources, both physical and managerial. This may starve other non-CPA-planned tasks to their detriment, and it may produce exceptionally good results on the "CPA" job which cannot be reproduced on later jobs. Whilst it is impossible to avoid this "halation" completely, its existence needs to be recognized.

(vii) CPA will involve committal to, and the acceptance of, responsibilities expressed in quantitative terms. Many supervisors find this difficult to accept, and will often try to escape by creating unreal problems. It is vital to make it quite clear that CPA is not a punitive device: it is a tool to assist, not a weapon to assault.

(viii) Difficulties in using CPA are *almost always* symptoms of some managerial weaknesses.

CHAPTER TWO

THE ELEMENTS OF A NETWORK—I GENERAL

As we have already seen, a project is represented by an ARROW DIAGRAM, which is not unlike the Work Study Engineer's Flow Chart. An arrow diagram is made up of only two basic elements—

(i) *An activity*, which is an element of the work entailed in the project. In some instances the "work" is not real in the sense that neither energy nor money is consumed, and in some cases (see dummy activities below) no time is used. However, ignoring these last cases, an activity is a task which must be carried out. Thus, "waiting delivery of component X" is an activity just as much as is "make component Y," since both are tasks which must be carried out. This "non-work" aspect of some activities is sometimes found difficult to accept until the test of *needfulness to the project* is applied. Once this test is applied it is clear that waiting for delivery is an activity in the sense in which the word is used in drawing networks.

(ii) *An event*, which is the start and/or finish of an activity or group of activities. The essential criterion is that a definite, unambiguous point in time can be isolated—a broad band of availability is of no use. The word "event" may be misleading here, since there may in fact be a concurrence of a number of separate events, and for this reason some authorities prefer the terms "node", "junction," "milestone" or "stage." In general, however, the word *event* is well established in the literature and, for this reason, will be used here, despite the possibility of misunderstanding.

One of the very useful by-products of drawing a network is that the diagram can be used as a means of communication. It can, for example, record decisions on how a project is to be completed, or it can enable an executive to pass on information to his successor or his subordinates. It is quite certain that the network will be seen and used by persons other than those who prepared it. With this "communicating" aspect in mind, it is important from a practical point of view that both events and activities should be *unequivocal* statements in *positive* terms which have *significance* within the context of the task being considered. Further, wherever possible, activities should be so chosen that the responsibility for carrying out the activity can be explicitly assigned.

Conventions Adopted in Drawing Networks

There are only two conventions usually adopted in drawing networks and, like all conventions, they may be ignored if circumstances warrant. In the early stages of network drawing, it is suggested that the conventions be respected until sufficient experience has been gained to justify dropping them. The conventions are—

(i) *Time flows from left to right.*

(ii) *Head events always have a number higher than that of the tail event.* This allows activities to be referred to uniquely by their tail and head event numbers, so that "activity 3 — 4" means only "the activity which starts from event 3 and proceeds to event 4"; it *cannot* mean "the activity which starts from event 4 and finishes at event 3". Some computer programs exist which do not require this convention to be followed, but experience will show that it is nevertheless a useful one, as discussed in the section on "looping" on page 19.

It may be convenient to remark here that it is not necessary for all numbers to be in sequence, that is, that numbers need not follow each other in natural order. In fact it is sometimes useful, when numbering events, to leave gaps in the normal sequence so that, if it is necessary to modify a drawing, it is not also necessary to renumber all events—a tedious task. This may then result in events being numbered 1, 2, 3, 7, 14, 15, 18, 19, 20 and so on, the numbers 4, 5, 6, 8, 9, 10, 11, 12, 13, 16, 17 not appearing at all. No inconvenience will be found to result from this. It is useful to realize that the head and tail numbers of the activities effectively specify the logic of the diagram, and that from a list of head and tail numbers, the network can be constructed.

The Graphical Representation of Events and Activities

Events are represented by numbers, usually within convenient geometrical shapes—often circles. Activities are represented by arrows, the arrow-heads being at the completion of the activities. The length and orientation of the arrow are of no significance whatsoever, being chosen only for convenience of drawing. The activity of leaving Place A and walking to Place B can equally well be represented by—

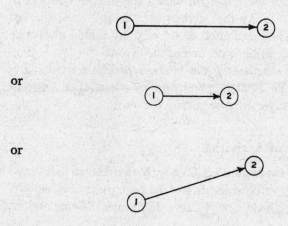

or

or

or

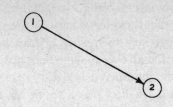

or

All of these have, within an arrow diagram, precisely the same significance, namely that to proceed from event 1 to event 2 it is necessary to carry out activity 1—2. It is equally not essential that arrows should be straight, although it will be found that the appearance of the whole diagram will be improved if the main portion of each arrow is both straight and parallel to the main horizontal axis of the paper on which the diagram is drawn. This will often require that arrows are "bent," as in the last sketch above. The description of the activity should always be written upon the straight portion of the arrow.

It is strongly recommended that wherever possible this method of drawing should be adopted. (*Note:* an example of a network drawn in this way can be seen on page 77.)

Identification of Activities

The event at the beginning of an activity is known as a "tail" or "preceding" event, whilst that at the conclusion of an activity is known as a "head" or "succeeding" event. Some writers refer to tail and head events as i and j events, this deriving from the generalization of an activity as—

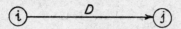

This usage is extremely convenient when drawing up tables, where the single letters i and j are simpler to use than the words "preceding" and "succeeding," as recommended in BS 4335: 1968, or "tail" and "head." All these ways of locating events will be found in the succeeding text.

Fundamental Properties of Events and Activities

Basically, the representation of events and activities is governed by one, simple *dependency rule* which requires that an activity which depends upon another activity is shown to emerge from the head event of the activity upon which it depends, and that only dependent activities are drawn in this way. Thus, if activity B depends upon activity A, then the two activities are drawn—

whilst if activity C is also dependent upon activity A, but is *not* dependent upon activity B, then the three activities are drawn—

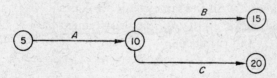

This dependency rule gives rise to two fundamental properties of events and activities—

(i) *An event cannot be said to occur* (or "be reached" or "be achieved") *until all activities leading into it are complete.* For example, if in a network the following (or its equivalent) is found—

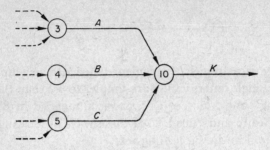

then event 10 can only be said to occur when activities 3—10, 4—10 and 5—10 are all complete.

(ii) *No activity can start until its tail event is reached.* Thus, in the following—

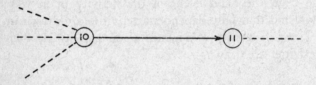

activity 10—11 cannot start until event 10 is reached.

These two statements can effectively be combined into a single comment, namely that "No activity may start until all previous activities in the same chain are complete." It must be understood, however, that this single statement has two facets as set out in (i) and (ii) above.

Two Errors in Logic

Two errors in logic may come about when drawing a network, particularly if it is a complicated one. These are known as *looping* and *dangling*.

(i) *LOOPING*

Consideration of any situation will show that the following must not occur—

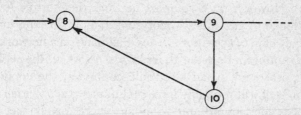

since this would represent an impossible situation: "event 10 depends on event 9 which depends on event 8 which depends on event 10 which depends on event 9 . . . ." If looping like this appears to arise, the logic underlying the diagram must be at fault, and the construction of the diagram must be re-examined. Adherence to convention (ii) above reveals the existence of a loop very easily.

(ii) *DANGLING*

Similarly, the situation represented by—

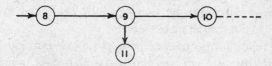

is equally at fault, since the activity represented by the dangling arrow 9—11 is undertaken with no result. Such arrows often result from hastily inserted afterthoughts. Two rules can be enunciated which, if followed, will avoid dangling arrows, namely, "all events, except the first and the last, must have at least one activity entering and one activity leaving them" and "all activities must start and finish with an event." There are special occasions when "dangling" activities can be accepted but any appearance of a "dangle" should be very carefully considered to ensure that it does not arise from an error in logic or an inadequate understanding of the task being considered. See "Multiple Starts and Finishes" below.

When set out in isolated form as above, both errors are quite obvious. However, in a complex network these errors (particularly looping) can arise, a loop for example forming over a very large chain of activities. Before finalizing on a network it is wise to examine for both the above logical errors. If the network is being processed by an electronic computer, the computer program itself will probably have preliminary tests written into it to check for looping and dangling before any further calculations are undertaken.

Multiple Starts and Finishes

There are occasions when a network can have more than one start and/or finish. These usually arise when some activities are entirely independent of the control of the user of the network. Wherever possible, these multiple starts/finishes should be "tied" in to the network by activity arrows giving the desirable, estimated or necessary time relationships between the "free" event and the rest of the network. The existence of these free events should, of course, be very carefully examined to ensure that they are not drawn incorrectly.

If time relationships cannot be specified or estimated, then the network can still be drawn and analysed, but the information derived for the free events or activities may well be limited. For example, it may be possible to specify only an earliest start time and not an earliest *and* latest start time for a free activity. It will be found helpful both to understanding and analysis if the "time flows from left to right" convention mentioned above is strictly adhered to when multiple starts/finishes occur.

"Merge" and "Burst" Nodes

Events into which a number of activities enter and one (or several) leave—

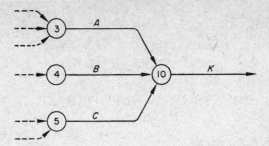

are known as "Merge" nodes. Events which have one (or several) entering activities generating a number of emerging activities are known as "Burst" nodes—

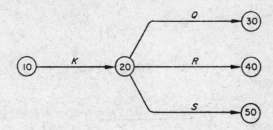

CHAPTER THREE

THE ELEMENTS OF A NETWORK—II DUMMIES

Dummy Activities

In some cases it is necessary to draw "dummy" activities, that is, activities which do not require either resources or time. These are usually drawn as broken arrows—

although some workers use a solid arrow with a subscript D, for dummy—

$$\xrightarrow{\quad\quad}$$
$$D$$

or a subscript O, indicating the need for zero resources and zero time—

$$\xrightarrow{\quad\quad}$$
$$O$$

However drawn, a dummy activity *is always subject to the basic dependency rule* that an activity emerging from the head of another activity depends upon that activity.

There are two occasions when dummies are used—

1. IDENTITY DUMMIES

When two or more parallel independent activities have the same head and tail events, the identity of the activities, as given by the event numbers, could be lost. For example, if two men leave place A to go to place B, one using one means of transport (i.e. resource), the other using another, this would be represented by—

where event 1 represents "leave A" and event 3 represents "arrive B." This would result in two activities having the same head and tail number. To avoid the resultant confusion a dummy activity is introduced, which can be either activity 1—2:

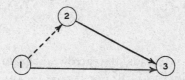

where both events 1 and 2 represent "leave A," or activity 2—3:

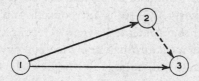

where both events 2 and 3 represent "arrive B." For the reason given in Chapter 15, "The Use of the Computer," the latter presentation is to be preferred.

2. LOGIC DUMMIES

When two chains have a common event yet they are in themselves wholly or partly independent of each other, then an error in logic could unwittingly arise. Continuing to examine men proceeding from one place to another, one man may leave place *A* to go to place *B* and leave there to progress onwards to place *C*, whilst a second man may leave place *K* to go to place *B*, collect an item deposited there by the first man, and then leave to progress onwards to place *L*. At first sight, the arrow diagram would be—

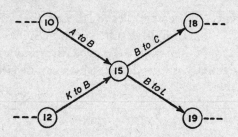

However, as it stands the arrow diagram indicates that neither activity 15—18 nor 15—19 can be undertaken until both activities 10—15 and 12—15 are complete. Clearly, this may not in practice be at all necessary: one may leave *B* to walk to *C* without waiting for the other man to finish his walk from *K* to *B*. Again, to resolve this difficulty a dummy activity 15—16 is introduced and, again, both events 15 and 16 represent "leave *B*." Of course, if it is *necessary* for the men to meet at *B* before proceeding onwards, then the original diagram will hold.

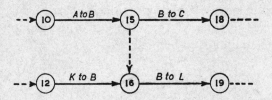

A very similar situation occurs when a pair (or more) of activities have one activity in common, and one activity *not* in common. For example, if activity L depends on activities A and B, and activity K depends only on activity A, the diagram may appear to be represented as in the first of the two drawings above, where—

activity 10—15 is activity A
" 12—15 " " B
" 15—18 " " K
" 15—19 " " L

Here again there is a dependence shown which is not in accordance with the actual situation; activity K is shown to be dependent on both activities A and B, whereas it is only dependent on activity A. Again, a dummy activity (15—16) is introduced, as in the second of the two drawings above, which "releases" activity K from its apparent dependence on A and B and shows it to be dependent only on activity A.

This second case of the need for dummies is found in practice to be a fruitful source of errors. It is highly desirable to examine any

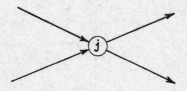

situation which occurs in the drawing of a diagram to ensure that the dependence of activities upon one another is quite clearly understood and represented.

The Direction of Dummies

Trouble is often encountered in assigning a direction to a dummy activity. If the purpose of the dummy is quite clearly understood, then the direction of the dummy becomes clearer.

Thus if, in the above, the dummy exists to release activity K from activity B then the dummy emerges from the tail event of activity K; on the other hand, if activity K depended on activities A and B and activity L depended only on activity B and *not* on activities A and B, the general configuration would remain unaltered except that the dummy arrow-head would point the other way, i.e. from event 16 to event 15. Reference should be made to an earlier discussion on Fundamental Properties of Events and Activities (page 17), and it should be clearly understood that the situation—

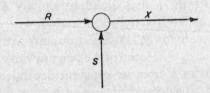

represents activity X being dependent on both activity R and activity S, whilst—

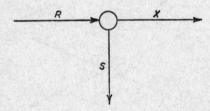

represents activity X being dependent only on activity R and *independent* of activity S.

The Location of Dummies

Difficulty is sometimes experienced in correctly locating dummies, a difficulty which diminishes with experience. A simple, albeit sometimes tedious, method of locating dummies, first introduced to your author by Norman Raby, is to draw *all* dependencies as dummies, and then to remove any redundant dummies. Consider the situation (assumed to be in the middle of a network) of the five dependencies—

Activity K depends on activity A
,, L ,, ,, activities A and B
,, M ,, ,, activities B and C

The six activities, complete with head and tail events, are drawn. If they can be roughly located in the correct position, it will be found helpful but not essential—

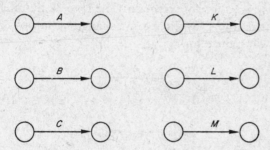

All five dependencies are drawn in as dummies—

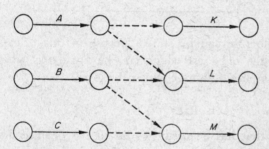

and the diagram then examined to see whether any of the dummies is unnecessary. For example, the dummy between A and K serves only to "extend" K, and its amalgamation into K would not change the situation so that—

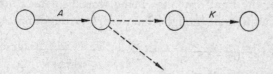

can become—

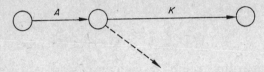

However, neither the *A* to *L*, *B* to *L* or *B* to *M* dummies can be removed without altering the logic of the situation, and they must, therefore, remain. The *C* to *M* dummy can be incorporated into *C* without any changes resulting, and the network will reduce to—

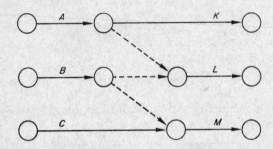

No simple rule can be stated for use in this "cleaning-up" process—the only test which can be applied is whether the absorption of a dummy changes the logic.

Overlapping Activities

In all that has been said so far, it has been assumed that activities are quite discrete, the start of a succeeding activity being delayed until a previous activity is complete. There are many occasions, however, when this is not so: a succeeding activity *can* start when the previous activity is only partly complete. For example, if a large number of drawings are being prepared, the final drawings being traced in ink from pencil sketches, it might be thought that the arrow diagram would be—

- - -①———DRAW PENCIL———→⑩———PREPARE TRACINGS———⑳- - -
            SKETCHES   FROM PENCIL
                                    SKETCHES

but this would indicate that *no* tracings could be prepared until *all* pencil sketches were complete. It may well be that tracing can start after some pencil sketches are complete, and that thereafter sketching and tracing will go on concurrently, sketches being fed through as they are completed to the tracers who will eventually finish tracing some time after the last sketch is received. This can be represented by breaking both activities into two fractions—

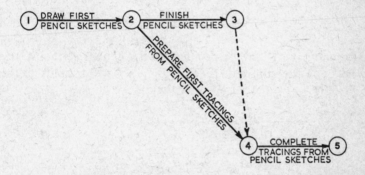

This shows that it is not possible to complete tracings until all the sketches are complete and the first tracings have been finished. The dummy 3—4 is used, of course, to enable the parallel activities 2—3 and 2—4 to be identified.

This "triangular" diagram represents the situation quite clearly but, when a number of these triangles are added in series, the drawing tends to become confused. A device which reduces the confusion is to draw the second activity arrow as a line "bent" at right-angles. Thus, if the two activities are P and S, "broken" into $P1$ and $P2$, $S1$ and $S2$, the diagram will become—

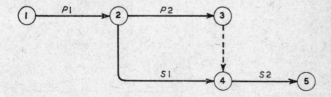

If it is now desired to add another concurrent activity R, this can be done by "breaking" $S1$ into two parts, $S1/1$ and $S1/2$, so that the diagram becomes—

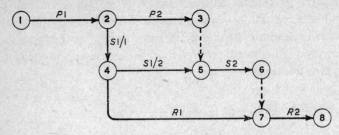

Alternatively, it may be decided that R cannot start until *the whole* of $S1$ is completed; thus the diagram will simplify to the following—

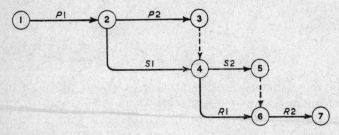

Note: Above, $R1$ depends on $P2$. If this is not so, and $R1$ can start with $S1$ complete and $P2$ incomplete, then a dummy between the junction of $S1$ and $R1$ and the junction of activity 3—4 and $S2$ will release the dependency. It will also remove the necessity for dummies 3—4 and 5—6. The result will be as follows—

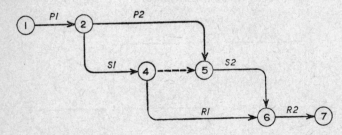

All these diagrams represent similar situations, namely a resultant activity starting before its originating activity is complete. Care must be exercised in using and analysing these networks, since some combinations of duration times can result in subsequent activities apparently being capable of finishing earlier than physically possible. To overcome this it is necessary *either* to examine the results of analyses and eliminate any impossibilities, *or* to impose a restriction on the "breaking" of the various activities—for example, ensuring that all concurrent activities are broken into the same fraction. Thus, if

$$P_1 = P_2, \text{ and } S_1 = S_2, \text{ and } R_1 = R_2$$

the physical meaning of the above diagram is: "When P is half-completed, S is started, the *second* half of S not starting until the *second* half of P is completed. The *first* half of R is only started when the *first* half of S is completed, and the *second* half of R is only started when the second half of S is completed." Of course, this may result (apparently) in tasks proceeding in a "jerky" manner, and the analyst must ensure that, if the activities are in fact necessarily continuous, the numerical analysis does not indicate a discontinuity. It is often most useful to break the base activities into three parts—for example, "start P", "continue P," "finish P."

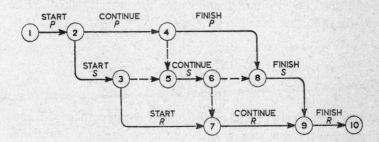

By adjusting the resources used on the activities, their duration times can be modified, and this, in conjunction with careful choice of the "start" and "finish" components, enables control to be exerted over the way resources are deployed.

CHAPTER FOUR

DRAWING THE NETWORK

POLICY is sometimes defined as the means whereby an objective is to be achieved, and in this sense an arrow diagram can be assumed to be a formal and explicit statement of policy. This concept will be found useful when considering the amount of detail that should be shown in an arrow diagram. BS 4335:1968 defines a *network* as "A diagram representing the activities and events of a network, their sequence and inter-relationships" and your author uses *network* synonymously with *arrow diagram*.

The Network as a Statement of Policy

In general, as one descends the hierarchy of an organization, the detail given in a policy statement increases, whilst its scope decreases. For example, the top management may decide to produce a new product X, and within the overall policy of the organization this might appear as—

The next level of management might be concerned with the deployment of the design and production resources, and the single arrow might become—

The design manager at the next level might then translate the single "Design Product X" arrow as a more complex network, of which a part might be as shown in the diagram overleaf.

The chief electrical engineer might then distribute the electrical designing among a number of different design engineers, so that in turn a single arrow at one level becomes a network at a lower level.

This idea of a network as a statement of policy is useful, and its implications are substantial. Policy denotes *known* objectives: it is not possible to lay down sensible policy unless the purpose for which it is in being is clearly known and explicitly stated.

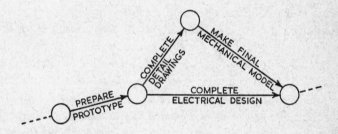

The first action, therefore, when drawing a network is to *define the purpose* of the project being considered. Stated thus baldly, the statement appears trite and, indeed, in the case of constructional and manufacturing enterprises, the objectives are usually very clearly known.

It is in design and development, an area where Critical Path Analysis can be very usefully employed, that the understanding of the task is often least well understood. Every development department must at some time have been faced with a request to design something which either has no target specification or else an extremely vague one. By agreeing very clearly with the "customer" what is required, enormous savings can be achieved.

It is extremely rarely that development work is undertaken at the frontiers of knowledge, and an unequivocal specification can be of inestimable value in allowing the designer to draw upon experience and in enabling him thus to make realistic plans.

The Network as a Budget

The Institute of Cost and Works Accountants in its invaluable *Terminology of Cost Accounting* defines a budget as follows—

"A financial and/or quantitative statement, prepared prior to a defined period of time, of the policy to be pursued during that period for the purpose of attaining a given objective."

This is a definition also of a network and it is often very helpful to consider the network as *a budget in terms of time*. The cost accountant has developed much skill in the assembly and use of financial budgets, and it is prudent to consider this experience when drawing and using networks. The author has found this parallel particularly valuable *when considering the detail which should be incorporated into a network*. A hierarchy of networks is just as appropriate as a hierarchy of budgets: it will enable problems to be identified without a mass of unnecessary detail, and it can locate responsibility at an appropriate point in the structure of the company.

Drawing the Network

When drawing an arrow diagram of a project, the major events and/or activities are fairly readily identified, and these should be approximately located in their correct positions relative to each other and the start and finish events on a large sheet of paper. Whilst it is possible at this stage to identify the events by numbers and prepare a list of events or activity descriptions to correspond to these numbers, it is always more convenient to write the descriptions on the diagram itself. For this reason it is unwise to try to make the arrows too small. Descriptions may well be abbreviated ("fdns" for "foundations," "cmpnts" for "components" . . .) and some organizations set up a glossary of approved abbreviations.

A supplementary list of activities, amplifying the descriptions used is often useful, but this is probably best prepared once the network has been tidied up and tested. Since duration times have significance only when methods and resources are defined, the most useful time to draw up this supplementary list is after duration times have been inserted, and the nodes numbered.

It is also unnecessary at this stage to try to make arrows straight, or always moving from left to right. Work on the diagram can proceed from both the start and the finish, and it is sometimes found that the project divides itself into a series of inter-related chains, and completing one chain at a time can be very helpful. The most useful pieces of equipment at this stage are a pencil and a good eraser: chalk and a blackboard are excellent alternatives.

The major events and activities having been drawn, the network can be completed by filling in the minor events and activities. A problem which repeatedly arises is to decide when to stop in writing down minor activities. If too much detail is written into a diagram it becomes excessively large, and the subsequent analysis increases in complexity without usefully increasing in value. There is, furthermore, the danger of losing sight of the physical realities underlying the diagram, and the whole analysis declining into a mathematical exercise. Rules on this do not seem to emerge, but it is suggested that three lines of inquiry can usefully be pursued—

(i) Can separate resources be shown with separate arrows?

(ii) Does any single arrow cover responsibilities assignable to more than one person?

(iii) Is the detail greater than that which is usable by the person employing the network?

It must be remembered, of course, that the definition of a resource, and the accountability for a responsibility, will apparently change with the level at which a plan is being made.

The first rough diagram may now be re-drawn, and the straightening-up and disposition of arrows checked in accordance with the conventions set up above. It should be realized that, as arrows are not vectors, their lengths and orientations are determined *only* by the convenience of drawing and the logic behind the project. The properties of events and activities set out in the previous chapter must be written into the diagram, but the location of an activity, by considering that which takes place *previously, concurrently* and *subsequently*, will be found very helpful. It is also wise to investigate the need for dummies, ensuring that where necessary they are inserted.

The author has found the following practice invaluable: once a network has been drawn, start from the final event and move up each activity arrow, asking the question—

What had to be done before this activity could take place?
Having reached the first event, move down each activity arrow asking the question—

What can now be done after the completion of this activity?
Carefully and systematically carried out, this procedure will be found to be of great assistance in ensuring logical cohesion.

As pointed out earlier the network is a statement of policy and, consequently, once the network is adopted, it commits the organization to a course of action, along with all the concomitant administrative procedures. Numbers of people— for example, departmental managers, site foremen, section leaders—are thus committed to *and will be held responsible for* carrying out tasks laid down in the network. If for no other reason, therefore, it is very wise indeed for the planner drawing the network to enlist the aid of the appropriate executives when drawing each part of the diagram. This may mean that the planner has to sit down separately and in conference with all the interested parties while individual responsibilities are being painfully worked out and agreed. It is often a temptation for a planner to try to carry out the whole operation by himself. This temptation should be resisted except in the case of very simple or frequently repeated tasks.

In order to locate responsibility and authority quite un-ambiguously, it is helpful to redraw the diagram so that it is divided horizontally into responsibility areas, and vertically into broad time areas. This will result in a diagram appearing as shown overleaf, and to ensure that responsibilities are under-stood *and accepted*, the signatures of the appropriate officers are required to appear on the diagram itself. This type of technique can prove of considerable benefit to the structure of the com-pany, and some of the results obtained will be—

(*a*) a clear understanding by all managers of the work they are committed to do;

(*b*) the delineation of responsibilities between managers;

(*c*) an investigation into the organization of, and pro-cedures used in, the company;

(*d*) the application of current experiences to the planning function.

Interfacing

When projects are particularly large or complex, it is sometimes desirable to construct a number of small networks based either upon resources or responsibilities. These can then be amalgamated into a larger complete network by means of the common events or activities. These are conventionally repre-sented either by double concentric circles or double parallel lines with a single arrow head, and they are known as *interface* elements, the amalgamating procedure being *interfacing*.

Duration Times

Once the logic behind a project is agreed, and the arrow diagram itself drawn, it can be completed by adding to each activity, as a subscript, its *duration time*. The duration time is the time which should be expended in carrying out the activity. It is not *necessarily* the time between the head and tail events; for

Design development	Body plan	Loft	Cut plates	Fabricate	Erect steel

Naval design — J Watson

Drawing preparation — M.O Flynn

Loft and material — R. Dupont

Production planning and control — A. Brown

Steel preparation and erection — G. Swan

Fitting out — G. Marlow

Engines — W Culverhouse

Horizontal divisions = areas of responsibility

Vertical divisions = sequence

REFINEMENTS TO A NETWORK

(By permission of International Computers & Tabulators Ltd.)

example, event 1 may take place in week 0, and event 2 in week 10, but the duration time of activity 1—2 may be only 4 weeks, in which case the activity is said to possess *float* (in this case 6 weeks). This matter will be discussed more thoroughly later.

As with all scheduling techniques, the times assigned to activities must be realistic; that is to say, they must take into account all local circumstances. Using the Work Study Officers' *Standard* Times is quite inappropriate here, since the actual work may not be performed "at standard." "Actual" or "observed" times are much more appropriate, although in many cases such times are not available. However, the principle is clear: the duration times need to be *realistic* rather than *desirable*, and they should be accepted by those held responsible for their achievement. Again, the experience of the Cost Accountant is pertinent: costs should be agreed with, and not imposed upon, the manager concerned. The *Cost Handbook* (Ronald Press Company, New York) contains the following (Section 20.20)—

". . . Primary responsibility for preparation of the budgets should rest with the supervisors of the various segments of the business. For example . . . the sales manager should participate actively in the development of the sales budget, since he will be the individual primarily responsible for the execution of the sales plan. This general procedure is equally applicable to every other segment of the business and should be vigorously pursued. . . .

"In this connection Francis (*Controller* Vol. 22) points out that:

'Budgets are frequently developed by one or two key individuals. . . . Sometimes management without the prior knowledge or approval of the operating executives in the (various) departments. This is the worst type of budget procedure and quickly defeats the objectives of forward planning.'"

Duration Times under Uncertainty

Not infrequently duration times are held to be impossible to estimate ("I've never designed one of these before. . . ." "We won't know what we have to do until we've taken it down . . ."). However, it is extremely rare to find a situation where it is not vital to carry out a task in a limited time: a design *must* be completed in such-and-such a time in order to allow the product to be sent to a customer/shown at an exhibition/submitted to test conditions/. . ., and equipment usually has to be repaired before a particular date in order to keep production going/allow the chairman to go on his holidays/avoid a power cut/. . . and so on.

In situations like these it may be useful to proceed as follows—

(i) Establish the *purpose* of the task as accurately, and in as much detail, as possible.

(ii) Establish any fixed dates. If no completion date is given, agree a reasonable target completion date.

(iii) Assign duration times wherever possible.

(iv) Break "uncertain" activities into smaller parts, for some of which times can be readily agreed since historical data are available.

(v) Examine the residual uncertain areas to see if there are any precedents which can act as guides.

(vi) Finally, assign to these uncertain activities as much time as possible without over-running the agreed or imposed target date. These activities can then be re-examined, asking the question—

"Can activity *X* be completed in time *t*?"

which is a more pointed and stimulating question than the original—

"How long will it take to complete activity *X*?"

Always separate the decisions on logic from the decisions on time.

PERT

In highly uncertain areas it may be better to use a "bracket" of times, giving an estimate of the "best" and "worst" and "most likely," and this is the situation for which PERT was originally designed.

PERT operates by assuming that the three-time estimates form part of a population obeying a β-distribution, so that the "best" (a), "worst" (b) and "most likely" (m) times can be compounded to give a single "expected time" (t_e) as follows—

$$t_e = \frac{a + b + 4m}{6}$$

and this t_e is the time used in all calculations. The choice of the β-distribution is not justifiable on experimental grounds, but it is computationally easy to handle, and its users state that it gives significantly useful answers.

A further use of the three-time estimate technique is to calculate the probability of any event occurring at any particular time. This manipulation has been very extensively discussed in the original PERT publications referred to in the Selected Readings. It is your author's experience that three-time estimates are occasionally useful to derive a single time estimate, but that the effort of carrying out the probability calculations is not repaid in any way by the value obtained from the resultant probabilities.

Assigning Duration Times

When assigning duration times throughout a project, an action which should be carried out as late as possible in the planning sequence, it is sometimes helpful to consider activities in a random sequence. The point here is that if, say, activities are considered sequentially in chains, it is possible that the duration time assigned to one activity might affect the choice of duration time for later activities. For example, if it is realized that a "long" activity might jeopardize the overall completion date, there is a temptation to "shrink" subsequent duration times to give an acceptable overall answer. As with any planning method, CPA is no more accurate than the information fed into it, and it is often very difficult to avoid unwittingly colouring estimates when the final answer can apparently be seen.

Numbering the Events

The final task in network drawing is to number the events in accordance with the convention set out on page 14. If the re-drawing of the network has, as recommended, caused all arrows to show time flowing from left to right, then numbering becomes quite simple. A straight edge is laid across the network, at right angles to its axis, drawn across the network from start to finish, the nodes being numbered as they are exposed.

Listing the Events and Activities

It is usually necessary to list the events and/or activities, giving descriptions and, in the case of activities, duration times. This task needs to be carried out systematically to avoid over-looking anything. Events are simple to list, but it will be found best in the case of activities to start with the first event number and list in sequence all activities starting with that event number. Then proceed to the next event number and clear all activities starting with *that* number, and so on throughout all the event numbers. Two checks are then possible—

(1) Every event (except the last) must have at least one activity emerging from it.

(2) There must be the same number of activities listed as there are arrows in the diagram.

It is at this stage that a definitive statement of resources, methods and limitations can best be made. It is often found helpful to produce a code list for each resource, for example:

A = Resource type 1, B = Resource type 2 . . . , so that the resource statements can be made very short ($3A$, $2B$ signifying 3 units of Resource type 1, 2 units of Resource type 2). When resource limitations are likely to occur, the resources required can usefully be entered upon the network itself—

Notes on Drawing Arrow Diagrams

1. Conventions to be used—
    (a) Time flows from left to right.
    (b) Tail events have a lower number than head events.

2. Identify objective.

3. Identify major events/activities.

4. Locate major events/activities on large sheet of paper.

5. Draw first diagram, joining major events by activity arrows and minor events/activities.

6. Check activity locations by asking—
    (i) What *has* happened?
    (ii) What *is* happening?
    (iii) What *will* happen?

7. For every activity ask—
    (i) What had to be done before this?
    (ii) What can be done now?

8. Check—
    (i) No looping.
    (ii) No dangling.
    (iii) All events are complete if all entering activities are complete.

9. Redraw diagram, numbering events in accordance with 1(*b*) above.

Remember: (*a*) Arrow lengths and orientations are not significant.

          (*b*) Events can be separated by dummy activities.

10. Avoid excessive detail.

11. Essential equipment: Pencil, eraser, and large sheets of paper.

CHAPTER FIVE

ANALYSING THE NETWORK: ISOLATING THE CRITICAL PATH

THE previous chapters have discussed how a network may be drawn, and the work so done is, in itself, invaluable. It has imposed a discipline upon the planners, forcing consideration of WHAT has to be done, WHEN, BY WHOM and IN WHAT TIME. It has also provided a clear, unambiguous statement of policy which is readily understood by all potential users. Thus, if no further action were taken, considerable benefits would already have been derived. However, it is possible, by using only the very simplest arithmetic, to extract a considerable amount of extra information.

Calculating the Total Project Time

The Total Project Time is the shortest time in which the project can be completed, and this is determined by a sequence (or sequences) of activities known as the Critical Path (or Paths). In order to calculate the Total Project Time, carry out a *forward pass*, that is—

(1) Start from the left of the arrow diagram (i.e. at the first event).

(2) Give the first event a time, o. *Note:* This is not equivalent to saying that all emergent activities must start at time o.

(3) Proceed to each in order and calculate the earliest possible time at which the event can occur. *Note:* If several

activities lead into an event, the earliest time is fixed by the *longest* chain. For example, in the following network—

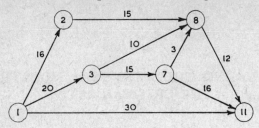

if event 1 is at week 0
then event 2 has an earliest time of 16 weeks.
Event 8 has three chains leading into it—
    (*a*) 1 —— 2 —— 8   (16 + 15 weeks)
    (*b*) 1 —— 3 —— 8   (20 + 10 weeks)
    (*c*) 1 —— 3 —— 7 —— 8 (20 + 15 + 3 weeks)
and the earliest time for event 8 is determined by the longest chain—in this case, chain (*c*) 1 —— 3 —— 7 —— 8 which has a combined duration time of 38 weeks. Hence the earliest time for event 8 is 38 weeks.

    (4) List these earliest event times—

Event No.	Earliest Time
1	0
2	16
3	20
7	35
8	38
11	51

Clearly, the total project time is given by the earliest time for the final event, which in the above example is event number 11. The total project time is thus 51 weeks.

Isolating the Critical Path

Continuing the analysis, the critical path can be isolated by carrying out a *backward pass*—

(5) Start now from the right (i.e. the last event).

(6) Give to this event its earliest time—in the example, week 51, that is, "turn round" at week 51. *Note:* This is not equivalent to saying that all entering activities must finish at week 51.

(7) By subtracting duration times, calculate the latest possible occurrence time for any event, assuming the final event is fixed at (6) above. By latest possible occurrence time is meant the latest possible time at which an event can take place without jeopardizing the total project time calculated above. As before, the latest event time is fixed by the longest chain leading into the event, remembering of course that calculations are being made from the right. For example, in the above diagram—

Event 11 has a latest time of 51 weeks

" 8 " " " " " (51 − 12) weeks
$$= 39 \text{ weeks}$$

Event 7 has two chains leading out of it—
(a) 7 —— 11
(b) 7 —— 8 —— 11

Latest time by chain (a) $= 51 - 16$ weeks $= 35$ weeks
" " " " (b) $= 39 - 3$ " $= 36$ "

Hence the latest time for event 7 is 35 weeks, since it is essential for both chains to be completed by time 51.

As a check on the accuracy of calculation of the event times, it should be noted that the difference between the earliest and latest event times for the final event must be the same as the difference between the earliest and latest event times for the first event. Thus when "turning round" at the earliest event time for the final event, the latest event time for the first event must be equal to the earliest event time for the first event.

(8) List the latest event times—

Event No.	Latest Time
1	0
2	24
3	20
7	35
8	39
11	51

(9) Combine the lists 4 and 8 above—

	Time	
Event No.	Earliest	Latest
1	0	0
2	16	24
3	20	20
7	35	35
8	38	39
11	51	51

Note: For event 1 and event 11, the earliest and latest times coincide.

(10) The critical path lies along those activities whose earliest and latest times for the tail events *and* for the head events are the same, and whose duration times are equal to the difference between the head and tail event times. (*Note:* Both tests must be applied.)

In the sample network above, the critical path lies along path—

$$1 \text{———} 3 \text{———} 7 \text{———} 11$$

In this case there is only one critical path; in other configurations, more than one path may emerge.

The Calculations in Practice

It is clearly not necessary to list separately the earliest and latest event times, since one combined list may be readily prepared. The two lists and the above detailed calculations are given here for ease of explanation. In practice it will invariably be found that, when not using a computer, the calculations are best carried out *on the diagram itself*. A number of techniques are available to do this: for example, marking the earliest and latest event times with $E = \dots \quad L = \dots$

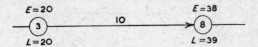

or enclosing the earliest and latest times within geometrical symbols, for example, a square for the earliest and a triangle for the latest event times.

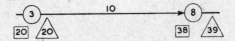

An ingenious technique, attributable to D. Whattingham of C.J.B. Ltd., is to replace the simple event circle by one divided into four segments, writing the event number in the bottom segment—

The earliest event time is then written in the left-hand segment, and the latest event time in the right-hand segment—

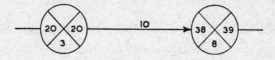

This technique is particularly useful when recording progress once the project is under way, when the actual occurrence time can be written into the top segment.

Conventionally the critical path is coloured red, or if this is not possible, is indicated by a pair of small, parallel lines set across each critical activity—

Using $E =$ $L =$ technique, the sample network would look like—

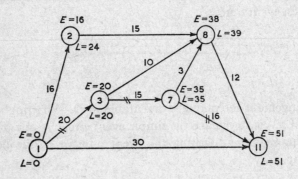

Activity Times

Since activities cannot start until their tail events are complete, and must not finish after their head events must start, the head and tail event times can be considered to fix boundaries between which activities can "move." It is possible to describe these "movements" by four simple times—

(1) *The earliest start time* is the earliest possible time at which an activity can start, and is given by the earliest time of the tail event. Thus, the earliest start time for activity 2—8 is the earliest time for event 2—that is, 16 weeks.

(2) *The earliest finish time* of an activity is the earliest possible time at which an activity can finish, and is given by

adding the duration time to the earliest start time; again, for activity 2—8 this is 16 + 15 weeks = 31 weeks.

(3) *The latest finish time* is found by taking the latest event time of the head event; again, for activity 2—8 this is the latest event time for event 8, that is, 39 weeks.

(4) *The latest start time* is the latest possible time by which an activity can start, and is given by subtracting the duration time from the latest finish time; for activity 2—8 the latest start time is 39 – 15 weeks = 24 weeks.

Summarizing the above—

Activity	Duration	Start Time		Finish Time	
		Earliest	Latest	Earliest	Latest
2—8	15	16	24	31	39

This can be done for all activities. The significance is that, considering activity 2—8, it must start between weeks 16 and 24 and must finish between weeks 31 and 39. An earlier start is impossible, whilst a later start will increase the overall performance time for the project; in fact it will shift the critical path from 1—3—7—11 to 1—2—8—11.

Note. Do not confuse *Event* times with *Activity* times. The earliest start time of an activity coincides with the earliest time of its tail event, and the latest finish time of an activity coincides with the latest time of its head event. However, the *latest* start time of an activity does not *necessarily* coincide with the latest time of its associated tail event, nor does the earliest finish time *necessarily* coincide with the earliest time of its head event; such coincidences only apply to activities on the critical path. Thus, the "Start Time *L*" and the "Finish Time *E*" in the following table *cannot* be read directly from the *E* = *L* = diagram, but *must* be derived from the "Finish Time *L*" and the "Start Time *E*." It may be helpful to consider that event times *E* look forward and event times *L* look backward.

The earliest and latest start and finish times for the whole of the sample network are—

Activity	Duration	Start Time		Finish Time	
		Earliest	Latest	Earliest	Latest
1— 2	16	0	8	16	24
1— 3	20	0	0	20	20
1—11	30	0	21	30	51
2— 8	15	16	24	31	39
3— 7	15	20	20	35	35
3— 8	10	20	29	30	39
7— 8	3	35	36	38	39
7—11	16	35	35	51	51
8—11	12	38	39	50	51

An Alternative Method of Calculating Earliest and Latest Event Times

There are a number of alternative methods of calculation, one of which follows. However, the author feels that it is important that the physical meaning underlying earliest and latest times should be thoroughly understood, and that this understanding is best obtained by carrying out the calculations as set forward previously.

This alternative method uses a simple matrix which is set up by drawing a square with two rows and two columns *more* than the number of events. Label the top left-hand corner E and the bottom left-hand corner L. From the left-hand corner of the second square in the first row, draw a diagonal through all the diagonal squares to the bottom right-hand corner of the last square in the penultimate row. Label the top half of the second square in the first row j and the bottom half i. Along the top row (i.e. opposite j) write all the event numbers, and down the second column (i.e. opposite i) also write all the event numbers.

For the simple example network already discussed, the square will appear as—

E	j \\ i	1	2	3	7	8	11
	1						
	2						
	3						
	7						
	8						
	11						
L							

Each small square is called a cell, and in the cell opposite heap and tail numbers (i.e. opposite i and j numbers) fill in the duration time of the corresponding activity. For example, in the second row (labelled 1) under the fourth column (labelled 2) fill in the duration time of activity 1—2, that is, 16 weeks, and in the same row under the fifth column (labelled 3) fill in the duration time of activity 1—3, that is, 20 weeks, and in the same row in the last column (labelled 11) the duration time of activity 1—11. Repeat this for all activities. The square will then look like—

E	j \\ i	1	2	3	7	8	11
	1		16	20			30
	2					15	
	3				15	10	
	7					3	16
	8						12
	11						
L							

The square is now in a form from which calculations can be made. To find the earliest event times proceed as follows:

Start at the first event in the i column and move right along the row until the diagonal is reached (i.e. until $j = 1$ is reached). Set in the E column the figure found above the diagonal. In this case it is o. (*Note:* This first step is unnecessary but is inserted to set the pattern for subsequent events.) Move now to the next event lower in the i column (in this case to $i = 2$) and again move to the diagonal. Set in the E column against $i = 2$ whatever number is found above the diagonal *plus* whatever is opposite that number in the E column—in this case $16 + 0 = 16$.

Proceed to $i = 3$ and $i = 7$ when $20 + 0$ and $15 + 20$ are entered in the E column. When $i = 8$, a slightly different situation arises since there are three numbers above the diagonal. The number which is entered in the E column is that number which is the greatest from the sum of the individual cells *plus* the corresponding E number. Thus, in this case there are three cells to be considered with values 15, 10 and 3 respectively, the corresponding E numbers being 16, 20 and 35, thus—

E	$i = 8$	$E + j$
16	15	31
20	10	30
35	3	38

The greatest sum is 38, and this is entered in the E column against $i = 8$. The same procedure is then repeated for $i = 11$, that is, the individual cells in the $j = 11$ column are added to the corresponding E column, and whichever sum is greatest is entered in the E column against $i = 11$, thus—

E	$i = 11$	$E + j$
o	30	30
35	16	51
38	12	50

so that 51 is entered against $i = 11$.

This whole process will fill the E column and give the earliest times for each event opposite the corresponding event number. To find the latest times for each event, the whole process is reversed, the latest times appearing in the L row at the bottom, starting from the bottom right-hand corner.

The latest time for the final event is, as previously, the same as its earliest time, and this is filled in in the bottom right-hand corner, that is 51 is filled in in the L row in the $j = 11$ column. Proceed to the next column and move upwards until the diagonal is reached, and then move along that row, subtracting the number in the row from the corresponding L number. In this case there is only one number in the row (12) and this is subtracted from the 51 in its own column. Thus, in the $j = 8$ column, a value of $L = 51 - 12 = 39$ is entered.

The same procedure applies for $j = 7$ except that in this case there are two numbers in the $i = 7$ row, namely 3 and 16 with corresponding L numbers of 39 and 51, and the smaller difference is entered thus—

$i = 7$	3	16
L	39	51
$L - i$	36	$\boxed{35}$

so that 35 is entered against $j = 7$. Similarly, for $j = 3$, proceed up the $j = 3$ column until the diagonal is reached; then proceed along the intersecting $i = 3$ row, subtracting the activity times from the previously recorded L times thus—

$i = 3$	15	10
L	35	39
$L - i$	$\boxed{20}$	29

so that an L figure of 20 is entered beneath $j = 3$. For $j = 2$, only one figure is possible $(39 - 15)$, whilst for $j = 1$, there are three possible figures—

$i = 1$	16	20	30
L	24	20	51
$L - i$	8	$\boxed{0}$	21

and a figure of o is entered under the $j = 1$ column. This provides a check on the whole calculation, since the earliest and latest times of the *first* event (like the final event) must be the same.

The completed matrix will appear as shown opposite; from this the earliest and latest times of the various events can be directly read.

One useful by-product of this method is that it gives a simple way of identifying those events which are directly linked to others; thus from the first line one can see that event 1 is directly linked to events 2, 3 and 11. The second row shows that event 2 is only linked to event 8, and so on. In a small network this identification is of little value, but in larger networks it can save much searching.

E	$\dfrac{j}{i}$	1	2	3	7	8	11
O	1		16	20			30
16	2					15	
20	3				15	10	
35	7					3	16
38	8						12
51	11						
L		O	24	20	35	39	51

CHAPTER SIX

ANALYSING THE NETWORK: FLOAT OR SLACK

As has already been described, the earliest tail and the latest head event times form boundaries within which activities are able to move.

Total Float

Looking again at activity 2—8, as shown in the following diagram—

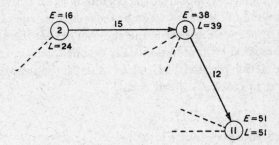

it will be seen that the earliest possible time the activity can start is week 16, whilst the latest possible time it can finish is week 39. Thus, it can be said that—

$$\text{the maximum available time} = 39 - 16 \text{ weeks}$$
$$= 23 \text{ weeks}.$$

Now the activity only "needs" the duration time in order that it can be completed, that is—

$$\text{the necessary time} = 15 \text{ weeks}.$$

Thus, the activity can "expand" or "move" by $(23 - 15) = 8$ weeks. Any expansion or movement *greater* than this will

57

change the critical path and increase the overall project time. This time of 8 weeks is known as the *total float* possessed by the activity.

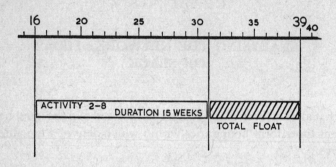

It must be realized that the total float is here shown as appearing as time at the *end* of an activity, but that this is not necessarily the case. Float can appear at the beginning of an activity, that is, the starting of the activity can be delayed after the tail event is reached; or it can appear *in* the activity, so that the duration time is increased beyond that initially planned; or it can appear after the activity is finished, while other activities are being concluded to reach the head event.

Free and Independent Float

Examining activity 8—11 in the same way it will be seen that—

the maximum available time	=	$(51 - 38)$ weeks
	=	13 weeks
the necessary time	=	12 "
hence, float	=	1 "

However, if activity 2—8 actually absorbs all its float of 8 weeks, event 8 will be reached by week $(16 + 15 + 8)$ = week 39. Thus activity 8—11 cannot possibly start until week 39, and the available time $\qquad$ = (51 – 39) weeks

$\qquad\qquad\qquad\qquad\qquad$ = 12 weeks

the necessary time $\qquad$ = 12 „

hence, float $\qquad\qquad$ = 0 „

so that, if activity 2—8 absorbs all its float, activity 8—11 has no float remaining. On the other hand, if activity 2—8 absorbs only 7 weeks or less of its float, the float in activity 8—11 remains unaltered at 1 week.

It can thus be said that activity 2—8 has 8 weeks' *Total* float, of which 7 can be used without reducing the float in any succeeding activity. One way of expressing this is to say that there is an Interference Float of 1 week associated with activity 2—8. A more common, and more useful, mode of expression is to say that activity 2—8 has a *Total* float of 8 weeks and a *Free* float of 7 weeks.

In the planning stage it may be decided to increase the duration time of activity 2—8 (for example, by reducing the resources allocated to it and thus increasing its performance time). If this is done, then the float available in *previous* activities will be reduced, so that the term *Free* indicates only that use of the float will not affect any succeeding activities. Cases do arise where the absorption of float affects neither earlier nor later activities, and the float is then said to be *Independent*.

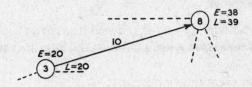

Activity 3—8 here has a maximum available time of (39 – 20) weeks = 19 weeks, and a necessary time of 10 weeks, so that the *Total* float is 9 weeks. Analysing as before, it will be found that the *Free* float is 8 weeks. If all associated activities take all the float possible, that is, if the tail event is reached as *late* and the

head event occurs as *early* as possible, then the time available to 3—8 is a minimum—

$$\text{the minimum time available} \;=\; (38 - 20) \text{ weeks}$$
$$=\; 18 \text{ weeks}$$
$$\text{and the necessary time} \qquad =\; 10 \quad \text{,,}$$

$$\text{hence, the } Independent \text{ float} \;=\; \underline{\underline{8}} \quad \text{,,}$$

Summarizing this, we have—

Activity	Duration	Start		Finish		Float		
		Early	Late	Early	Late	T	F	I
3—8	10	20	29	30	39	9	8	8

As a further example of the different types of float, consider the following, which is part of a network not hitherto considered—

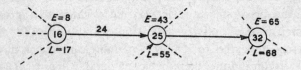

For activity 16—25:

$$\text{Maximum available time} \;=\; 47 \text{ weeks}$$
$$\text{Necessary time} \qquad\quad =\; 24 \quad \text{,,}$$

$$\therefore \text{ Total float} \qquad\qquad =\; \underline{\underline{23}} \quad \text{,,}$$

Float in activity 25—32 is reduced if event 25 cannot occur until some time *after* week 43. Thus, the free float in activity 16—25 is given by—

Earliest event time for event 16
+ duration time for activity 16—25
+ free float for activity 16—25
= earliest event time for event 25,

i.e. $8 + 24 + (\text{free float})_{16—25} \;=\; 43$

i.e. $(\text{free float})_{16—25} \qquad =\; 43 - 32 \text{ weeks}$
$$=\; 11 \text{ weeks.}$$

Minimum available time   =   26 weeks
Necessary time   =   24 „
                                 ────
Independent float   =   2 „
                                 ════

which, summarized as before, gives—

Activity	Duration	Start		Finish		Float		
		Early	Late	Early	Late	T	F	l
16—25	24	8	31	32	55	23	11	2

Negative Float

It is sometimes convenient to compare the overall project time with a target or acceptable time, and this can be very conveniently done by "turning round" at this target time. Thus, if the $E = L =$ technique is used, the target time is inserted at the final $L =$ figure. The latest event times are then calculated from this final $L =$ time, and float is again extracted. If the target time is greater than the total project time, then *all* activities will have positive float, whilst if the target time is *less* than the total project time, the critical path, and possibly some other activities will have *negative* float. This negative float is the time by which its associated activity must be reduced for the project to meet the target time. This extended concept of float then gives a precise definition of the critical path—

> The critical path in a network is that path which has least float.

Significance of Float

The importance of knowing the types of float depends upon the use made of the information. For example, if it is desired to reduce the effort on a non-critical activity, thus increasing its duration time but releasing effort for use elsewhere, then Independent float can be used without replanning any other activities. On the other hand, Free float can be used without affecting subsequent activities, whilst Total float may affect both previous and subsequent activities. Negative float indicates the reduction in duration time required to meet a target date.

Slack

A different expression of the ability of activities to move is given by considering the head and tail events. These have "earliest" and "latest" times, and slack is the difference between these times. Thus, for event 2—

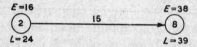

where the earliest event time is 16 weeks and the latest is 24 weeks, the slack is said to be 8 weeks, and for event 8 it is 1 week. Remembering that the beginning of an activity is the tail and the end is the head, it can be said that activity 2—8 has a tail slack of 8 weeks and a head slack of 1 week. Using the tail slack of an activity affects in general the slack in both earlier and later events, whilst using the head slack affects in general only the subsequent events.

The relationship between float and slack is—

Free float   = Total float – head slack
Independent float   = Free float – tail slack

Applying this to activity 2—8 we have—

Total float   = 39 – 16 – 15 weeks = 8 weeks
Head slack   = 1 week
Tail slack   = 8 weeks
∴ Free float   = Total float – head slack
                      = (8 – 1) weeks
                      = 7 weeks

Independent float = Free float – tail slack = 7 – 8 weeks = – 1 week which for practical purposes is taken to be zero.

As with float, slack can be a measure of the acceptability of the project as planned. Thus, if the critical path length is 51 weeks and the maximum acceptable time is 41 weeks, then events on the critical path are said to have −10 weeks' slack, or 10 weeks' negative slack. This usage is quite convenient when a project is actually running and slack can be calculated from actual rather than predicted duration times. Should slack be positive then it is possible to meet the accepted overall time without replanning. If slack is negative replanning is essential to return the project to its previously agreed overall time.

Float: A Summary

Independent Float: The time by which an activity can expand without affecting any other either previous or subsequent.

Free Float: The time by which an activity can expand without affecting subsequent activities. If it is absorbed at the planning stage the float in earlier activities will be reduced. Once a project is under way, the free float in an activity can be used once the tail event is reached without affecting any other activity in the network.

Total Float: The time by which an activity can expand. When total float is absorbed at the planning stage, the floats in both previous and subsequent activities may be reduced.

Negative Float: The time by which an activity must be reduced for the project to meet a target date.

Float and the Utilization of Resources

The statement that an activity possesses float is equivalent to a statement that the resources available for the performance of that activity are not fully used. Thus, the activity 2—8 already observed, which can "float" for 8 weeks between weeks 16 and 39, must have resources *available* for the whole of that time—that is, for 23 weeks—which are only *used* for the duration time of 15 weeks. Thus, the activity is only utilized for $\frac{15}{23}$ x 100 per cent of its possible utilization, i.e. it is only 65·2 per cent utilized.

In order to increase the utilization of the 2—8 resource, some of it may be transferred elsewhere, leaving only the essential 65·2 per cent resource on activity 2—8. If this is done, the duration time for activity 2—8 will perforce increase to 23

A GRAPHICAL SUMMARY

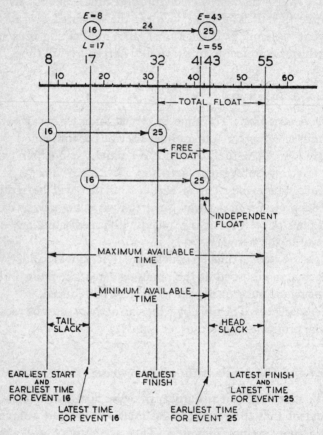

weeks, and the float will consequently disappear. Once this happens, of course, the critical path will lie through activity 2—8. In the example, if 2—8 increases *exactly* to 23 weeks, there will be two critical paths, namely

1—2—8—11   and   1—3—7—11

or, if for some reason 2—8 increases beyond a duration time of 23 weeks, then the critical path will shift from

1—3—7—11 to 1—2—8—11.

If *all* resources are fully utilized, the whole network becomes critical. At first sight this might appear to be a highly desirable situation in that there are no idle resources. It must be remembered, however, that idle resources represent some degree of flexibility in a project, and that to remove this flexibility might result in a state of crisis which could have been avoided—or at least alleviated—if some float had been available.

Thus, we have two general comments—

(1) Float represents under-utilized resources.
(2) Float represents flexibility.

Rules for Calculating Float

Total Float: Subtract the earliest time for the preceding event from the latest time for the succeeding event, and from this difference subtract the duration time.

Example: for activity 3—8 (page 46)

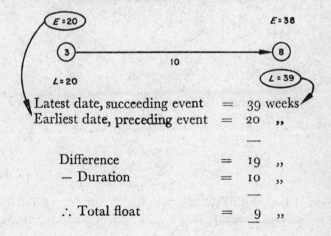

Latest date, succeeding event	=	39 weeks
Earliest date, preceding event	=	20 ,,
Difference	=	19 ,,
— Duration	=	10 ,,
∴ Total float	=	9 ,,

Note. When setting out an analysis in tabular form a useful alternative statement of the above rule is given by the BS 4335: 1968 definition of Total Float.

Total float is "Latest start date of activity minus earliest start date of activity. (May be negative.)"

Thus for activity 3—8

Latest start date   = $(39 - 10)$ weeks   =   29 weeks

Earliest   „ „   =   20   „

∴ Total float   = $\underline{\underline{9}}$ „

Free Float : here again the BS 4335: 1968 definition is useful— Free Float is "Earliest date of succeeding event minus earliest finish date of activity."

Example: for activity 3—8

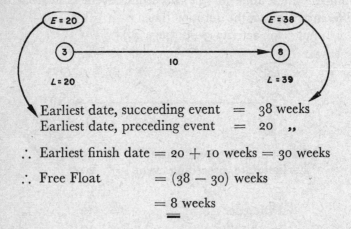

Earliest date, succeeding event   =   38 weeks
Earliest date, preceding event   =   20   „

∴ Earliest finish date = $20 + 10$ weeks = 30 weeks

∴ Free Float   = $(38 - 30)$ weeks

             = $\underline{\underline{8 \text{ weeks}}}$

Note. It is very difficult to derive free float from a tabulation, and it is always necessary to use the $E = L =$ analysed network.

The relationship between float and slack—page 62—gives an alternative method of calculation useful when the total float is known.

Example: for activity 3—8

Total float	=	9 weeks
− Head slack (39 −38 weeks) =	1	,,
∴ Free float	=	8 ,,

Independent Float: this is defined in BS 4335:1968 as—

Independent float is "Earliest date of succeeding event minus latest date of preceding event minus activity duration." (If negative, the independent float is taken as zero.)

Example: for activity 3—8

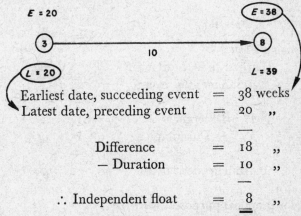

Earliest date, succeeding event	=	38 weeks
Latest date, preceding event	=	20 ,,
Difference	=	18 ,,
− Duration	=	10 ,,
∴ Independent float	=	8 ,,

Note. As with free float, calculation of independent float requires reference to the $E = L =$ analysed network. Again, the relationship between float and slack—page 53—gives an alternative method of calculation useful when free float is known.

Example: for activity 3—8

Free float	=	8 weeks
− Tail slack = (20-20) weeks =	0	,,
∴ Independent float	=	8 ,,

Further Illustration of Float Calculations

Activity 2—8

Total Float:

Latest start time	=	24 weeks
Earliest ,, ,,	=	16 ,,
∴ Total float	=	$\underline{\underline{8}}$,,

Free Float:

Earliest time head event	=	38 weeks
,, ,, tail event	=	16 ,,
Difference	=	22 ,,
— Duration	=	15 ,,
∴ Free float	=	$\underline{\underline{7}}$,,

Independent Float:

Earliest time, head event	=	38 weeks
Latest time, tail event	=	24 ,,
Difference	=	14 ,,
— Duration	=	15 ,,
∴ Independent float	=	$\underline{\underline{-1}}$,,

which for practical purposes is written

Independent float = 0 weeks

Alternatively

Free float	=	7 weeks
— Tail slack = (24 — 16) weeks =		8 ,,
∴ Independent float	=	$\underline{\underline{-1}}$,,

which is written as

Independent float = 0 weeks.

Activity 8—11

Total float = $(51-38)$ — 12 weeks   = 1 week

Free float = Total float — head slack = 1–0 = 1 week

Independent float = Free float — tail slack = 1–1   = 0

Generalized Rules for Analysis

Expressing the rules generally in mathematical terms we have, for the generalized activity i—j of duration D

    where the tail event of the activity is i

    and „ head „ „ „   „   „ j

and the earliest and latest event times are denoted by subscripts E and L

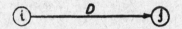

The earliest time of tail event i $= i_E$

„   latest   „ „   „   „ i $= i_L$

„   earliest   „ „   head „ j $= j_E$

„   latest   „ „   „   „ i $= j_L$

Then for activity i—j

the earliest start time $= i_E$

„ latest   „   „ $= j_L - D$

„ earliest finish time $= i_E + D$

„ latest   „   „ $= j_L$

„ total float $= j_L - i_E - D$

„ free float $= j_E - i_E - D$

„ independent float $= j_E - i_L - D$

and event slack for event i $= i_L - i_E$

Using the above rules for the sample network already discussed,

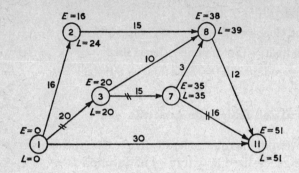

we have the following—

Activity	Duration	Start Time		Finish Time			Float		
		Early	Late	Early	Late	Tot.	Free	Ind.
1— 2	16	0	8	16	24	8	0	0
1— 3	20	0	0	20	20	0	0	0
1—11	30	0	21	30	51	21	21	21
2— 8	15	16	24	31	39	8	7	0
3— 7	15	20	20	35	35	0	0	0
3— 8	10	20	29	30	39	9	8	8
7— 8	3	35	36	38	39	1	0	0
7—11	16	35	35	51	51	0	0	0
8—11	12	38	39	50	51	1	1	0

Completing an Analysis Table

In practice it will be found most convenient, when compiling an analysis such as the above to proceed as follows—

(1) Fill in activity numbers, ensuring that every arrow in the network is represented by a head and tail number.

(2) Fill in duration times.

(3) Write down start time *Early* from the analysed network, using the *E* of the tail event.

(4) Write down finish time *Late* also from the network, using the *L* of the head event.

(5) Calculate start time *Late* from above by subtracting duration time from finish time *Late*.

(6) Calculate finish time *Early* from above by adding duration time to start time *Early*.

(7) Calculate total float by subtracting finish time *Early* from finish time *Late*.

(8) Calculate free and independent float from network using rules on pages 65 to 67 or relationship on page 62.

Verification of an Analysis

When an analysis has been made there are a number of simple checks that can be carried out which, though not in themselves conclusive, can bring some arithmetical and recording errors to light. For example—

(1) All activities with the same tail number have the same earliest start time.

(2) All activities with the same head number have the same latest finish time.

(3) Earliest start times are never larger than latest start times.

(4) Earliest finish times are never larger than latest finish times.

(5) Start times are always earlier than corresponding finish times.

(6) Free float can never exceed total float.

(7) Independent float can never exceed free float.

(8) Total float is the difference between earliest and latest start or finish times.

Intermediate Scheduled Times

Circumstances sometimes require that events other than, or as well as, the final event should take place at particular times. Should this be so, then these "scheduled" times can be inserted into the network by means of a solid arrow-head, surmounted by the scheduled time. For example—

indicates that event 75 must take place by time 100, that is, that all activities with a head number 75 must be complete by time 100. An arrow-head inverted below the event number indicates that activities with tail number 75 cannot start until time 100.

The scheduled time having been inserted, analyses can now be carried out (*i*) between the beginning and end of the network and (*ii*) between the beginning and/or end and the intermediate scheduled points, treating these points as if they were starts or finishes. This can then give rise to critical paths other than the main critical path, and these are known as secondary, tertiary, and so on, each with its own set of floats. Once such an analysis has been carried out, the meanings of the various factors will become immediately obvious.

In order that any reader who so desires can practise analysing networks, three simple small networks are shown below, with their analyses shown on pages 197 to 201.

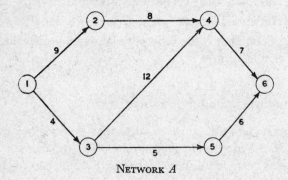

NETWORK *A*

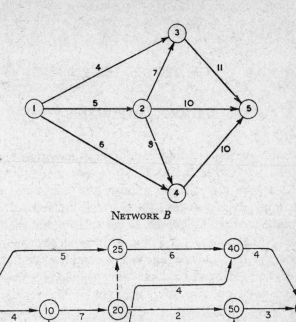

NETWORK B

NETWORK C

CHAPTER SEVEN

LADDER ACTIVITIES

A METHOD of representing overlapping activities, sponsored and used by ICL is the so-called "*Ladder* method." In this, *restraint* arrows are used to indicate the minimum intervals which must elapse between the start of one activity and the start of the next, and the finish of one activity and the finish of the next. For example, if P has a duration time of 3, and S may start when 1 unit of time of P is complete, and at least 2 units of time of S must elapse after the completion of P, S having a total duration of 5, then the ladder representation would be—

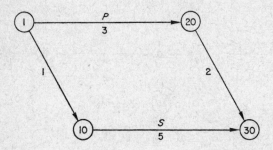

the arrows 1–10 and 20–30 representing the restraints imposed upon S and P. These constraints are sometimes known as "lead" and "lag" activities and are labelled accordingly, but it is helpful to indicate their physical meaning by labelling them "START . . ." and "FINISH . . ." As the computational methods for ladders differ slightly from conventional CPA, it is well to use a different node symbol as a "warning flag."

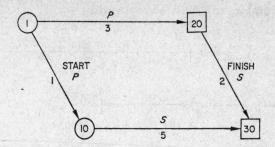

Ladders of this sort can be built up as the situation demands.

The different computational needs also require an "isolation" of the ladder from the rest of the network. This isolation generates two special "ladder" rules—

(i) a ladder activity, and its associated "start" activity must have a common tail node, and no other activity may emerge from that node;

(ii) a ladder activity, and its associated "finish" activity must have a common head node, and no other activity may enter that node.

Thus, the diagram—

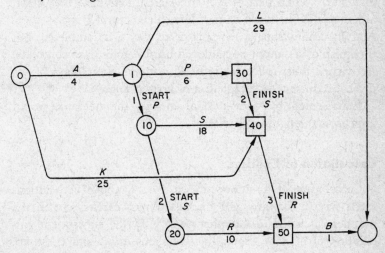

is unacceptable, and must be replaced by—

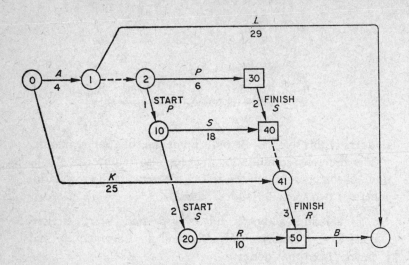

Note. The above figure represents a highly complex situation—

The opening of the project permits two activities A and K to start. After activity A is completed, two activities L and P may start. When part of P is completed, activity S may start, although the finish of S cannot be achieved until P is complete. Activity R may start when part of activity S is completed, but the finish of R cannot be achieved until K and S are complete. Activity B must follow activity R, and the completion of the project requires the completion of both B and L.

Represented by conventional means, the network would appear as the figure opposite.

Calculation of Ladders

Calculation of a ladder drawn to the above conventions starts with a forward and backward pass carried out in the manner discussed in Chapter 5. It is in the subsequent calculation of activity times that problems arise, and these are

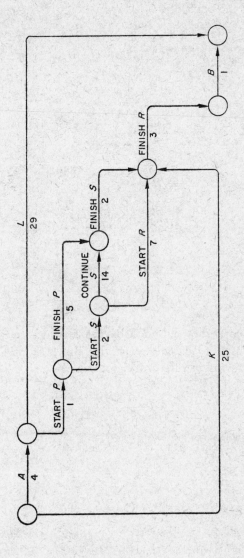

due to the special meaning of the restraint activities. To understand the problem it is useful to return to the physical meaning of the parts of the ladder.

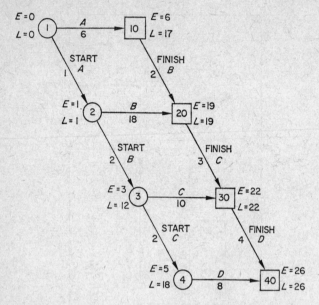

If the event times obtained by the forward and backward passes are used to calculate activity times as in Chapter 5, then the following results will be obtained—

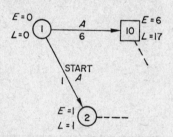

Activity			Start		Finish	
Numbers	Description	Duration	Early	Late	Early	Late
1—10	A	6	0	11	6	17
1—2	START A	1	0	0	1	1

so that apparently "START A" would finish at time 1, whilst A could start at time 11, that is, the whole activity could start after a part of itself had finished, something which is difficult to envisage.

To understand the physical nature of the situation, consider the Gantt Chart for the network in the "early" position—

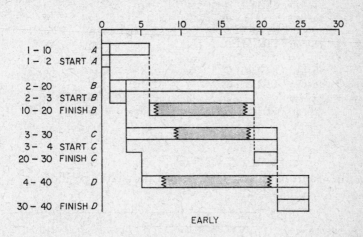

EARLY

Activities 1—10 and 1—2 must have a common start since they are inseparably joined, as are activities 2—20 and 2—3, and 3—30 and 3—4. Similarly the finishes of activities 2—20 and 10—20 must be the same, as must be 3—30 and 20—30, and 4—40 and 30—40. If activity 10—20 is now considered, it will be seen that it can *start* after A has finished, that is, at time 6, but that it *must* finish at the same time as B that is, at time 19. The duration time of activity 10—20 is only 2 units of time, so that there must be some "dead" time of magnitude $(19—6—2) = 11$ units of time incorporated into activity 10—20. This waste time (or "enforced idle time") is shown by the "dotted" shading and the ragged ends to the activity 10—20 bar.

In the same way, C must incorporate some enforced idle time (E.I.T.) in the early position, this being of magnitude

$(22—3—10) = 9$ units of time. D will also have E.I.T. of $(26—5—18) = 13$ units of time. The above arguments may be summarized in the following table.

Activity Numbers	Description	Duration	Start	Finish	E.I.T.
1—10	A	6	0	6	0
1—2	START A	1	0	1	0
2—20	B	18	1	19	0
2—3	START B	2	1	3	0
10—20	FINISH B	2	6	19	11
3—30	C	10	3	22	9
3—4	START C	2	3	5	0
20—30	FINISH C	3	19	22	0
4—40	D	8	5	26	13
30—40	FINISH D	4	22	26	0

TABLE I

If the Gantt chart is now drawn in the "late" position—

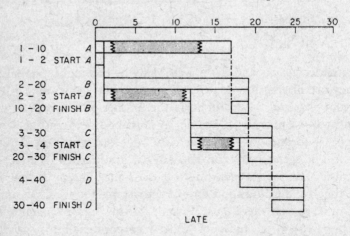

it can be seen, by a similar set of arguments that A may now have an enforced idle time of 11 units of time, its start being located by the start of the whole project, whilst its finish may "drift" to a position located by "FINISH B." "START B"

may have an E.I.T. of 9 units of time and "START C" an
E.I.T. of 4 units of time. The "late" position can be summarized
as follows.

Activity Numbers	Description	Duration	Start	Finish	E.I.T.
1—10	A	6	0	17	11
1—2	START A	1	0	1	0
2—20	B	18	1	19	0
2—3	START B	2	1	12	9
10—20	FINISH B	2	17	19	0
3—30	C	10	12	22	0
3—4	START C	2	12	18	4
20—30	FINISH C	3	19	22	0
4—40	D	8	18	26	0
30—40	FINISH D	4	22	26	0

TABLE II

Examination of Tables I and II in conjunction with the
forward and backward pass will reveal that the earliest start
and finish times for any activity are given by the E values at
the tail and the head of the activity, and that the latest start
and finish times are the corresponding L values. In both cases
the E.I.T. is given by subtracting the duration time of the
activity from the difference between the start and finish times.
Amalgamating Tables I and II into one Table, the following
is obtained—

Activity Numbers	Description	Duration	Early Start	Finish	E.I.T.	Late Start	Finish	E.I.T.
1—10	A	6	0	6	0	0	17	11
1—2	START A	1	0	1	0	0	1	0
2—20	B	18	1	19	0	1	19	0
2—3	START B	2	1	3	0	1	12	9
10—20	FINISH B	2	6	19	11	17	19	0
3—30	C	10	3	22	9	12	22	0
3—4	START C	2	3	5	0	12	18	4
20—30	FINISH C	3	19	22	0	19	22	0
4—40	D	8	5	26	13	18	26	0
30—40	FINISH D	4	22	26	0	22	26	0

The critical path is the path which has minimum float, or in this case, the special float defined as enforced idle time, and both the "Early" and the "Late" situations need to be examined. In this case the critical path is 1—2—20—30—40. Note that in the special case of ladders, using the isolating conventions above, the critical path is defined as that path which lies between events where earliest and latest times are equal.

The Ladder in Practice

The above discussion is unreal in the sense that it is unlikely that a task will be planned to have an enforced idle time. In practice, enforced idle time having been identified, the resources on the activities concerned will, if possible be adjusted so that the duration time of the activity will increase to absorb the idle time. Thus, if the "earliest" situation is acceptable, then attempts are likely to be made to increase, for example, C from a duration time of 10 to a duration time of 19 by diluting the resources applied. This will have the effect of causing C to become critical and as usual, the more efficient the deployment of resources, the less flexibility is available.

Note. The ladder convention is not easy to use, particularly in the absence of resource allocation and modification facilities. Used with care, however, it can provide a useful short-hand method of dealing with a difficult situation as in the diagram on page 77.

CHAPTER EIGHT

REDUCING THE PROJECT TIME

As already stressed, a network is a statement of policy, that is, a statement of the means whereby an objective is to be obtained. It is extremely rarely that only one acceptable policy can be formulated; further, it is equally true that almost any policy can be improved. This is very clearly recognized by the Work Study Engineer, whose basic tenet of faith is "There is always a better way."

This being so, it is desirable that when a network has been drawn it should be very carefully and critically examined. In complex projects the difficulty in examining all activities is so great that frequently no examination takes place at all, or alternatively only those activities in which the examiner has a particular interest are inspected. Critical Path Analysis has the tremendous advantage that, by isolating the *critical* activities, examination can be directed towards those areas which most significantly affect the overall time. In reducing the times of the initially critical activities, new critical paths may be created which, in turn, must be scrutinized.

The Questioning Method

In order that the examinations of the various activities shall be consistently useful, it is desirable to employ the well-tried Work Study technique of systematic questioning. In this, a number of questions are set up, and these questions asked of every activity. By asking the same questions in this apparently rigid way it is possible to ensure that a thorough examination is made of *all* alternatives. For a full discussion of this method, reference should be made to any one of the many textbooks on Work Study. The following should be considered only as an introduction to the method.

Activities can be considered to be of two kinds—

"DO" activities, where time is consumed in a task which, in itself, advances the project; and

"ANCILLARY" activities, where time is consumed in tasks which support "DO" activities.

For example, if a project involves the making of a component, the act of making is a "DO" activity, and the acts involved in setting up and breaking down the plant to carry out the making are "ANCILLARY" activities. Clearly, it is the "DO" activities which should be examined first, since if they can be reduced or eliminated, the "ANCILLARY" activities may either vanish or be reduced. (The author recalls a project which involved an activity "Assemble refrigerated tank," along with its associated activities of "Place orders," "obtain material," "test lagging" and so on. Discussion had centred on the problems of reducing purchasing time, until the "DO" activity— "assemble refrigerated tank"—was examined, when it was discovered that in fact this was an entirely unnecessary activity, and with its elimination, the ANCILLARY activities disappeared.)

Once these "DO" items on the critical path have been identified, they can be tested against a series of questions, which are dealt with more fully in R. M. Currie's *Work Study*.

(1) *Purpose* What is being done?
            Why is it being done?
            What else could be done?
            What should be done?

(2) *Place*   Where is it being done?
            Why there?
            Where else could it be done?
            Where should it be done?

(3) *Sequence*   When is it done?
                   Why then?
                   When else could it be done?
                   When should it be done?

(4) *Person*   Who does it?
                   Why that person?
                   Who else might do it?
                   Who should do it?

(5) *Means*   How is it done?
                   Why that way?
                   How else can it be done?
                   How should it be done?

Should *these* questions, having been applied to the critical activities, not produce the desired result, one other question may be asked, although it must be stressed that it is a dangerous one to ask, namely—

(6) Can an activity be reduced by increasing the *Risk*?

For example, it may be that the initial network has in it a "testing," "checking" or "proving" activity. Such activities can often be reduced, but with an increase in the risk of failure; thus, after drawings are completed, checking is often carried out, and if this is thorough then the checking time can be great—a substantial part of the initial drawing time. If this checking time is reduced, there is a greater chance that errors will slip through, with all the consequent undesirable results. To reduce this time, therefore, will increase the risk, and this decision must be squarely put to management for acceptance or rejection.

Reduction Involving Transference of Resources

The non-critical activities in a network can sometimes be used to obtain resources which can be applied to critical activities to reduce their durations. This is sometimes known as "trading off" resources.

Reduction Involving Increased Cost

All other methods having failed, a reduction in time may have to be obtained at an increased cost, usually by increasing the resources which are employed. If the costs to reduce times are known, then a table can be set up showing the relative costs for the reduction in time of each activity by the same amount. The cost incurred in reducing duration time by unit time may be defined as the "cost slope" thus—

for activity 1—3

Normal duration time of 20 weeks costs £200
and reduced „   „   „ 19   „   „   £220

hence the cost slope = £20 per week.

For the sample network already considered, the table of costs slope might be—

Activity	Duration	Total Float	Cost slope in £/week
1— 2	16	8	30
1— 3	20	0	20
1—11	30	21	10
2— 8	15	8	60
3— 7	15	0	45
3— 8	10	9	120
7— 8	3	1	10
7—11	16	0	15
8—11	12	1	95

Clearly, of the critical activities, activity 7—11 has the smallest cost slope, and it is desirable to investigate the practicability of reducing it first. These investigations may show not only that it can be reduced by 1 week, at an increased cost of £15, but further reductions are also readily obtainable. Inspection of Total Float, however, shows that two activities (7—8 and

8—11) will become critical if activity 7—11 is reduced by 1 week and, in fact, two critical paths

$$1—3—7—11 \text{ and } 1—3—7—8—11$$

will be formed. Thus, in order to reduce total project time, *either* the common part of these two paths (i.e. 1—3—7) must be reduced *or* the two branches (7—8—11 and 7—11) must be reduced simultaneously. The least expense is incurred when reducing the two branches by shrinking activity 7—8 (cost slope £10/week) and activity 7—11 (cost slope £15/week). This will produce an effective cost slope of £25/week, which is greater than the cost slope of activity 1—3 (£20/week), so that it would probably be desirable to investigate the reduction of activity 1—3 first.

Just as it is possible to reduce duration times by increasing costs, so may it be possible to reduce costs by increasing times. For example, activity 1—11 as planned has a duration time of 30 weeks and a Total Float of 21 weeks. Examination of activity 1—11 may show that its duration time could be increased to, say, 40 weeks while at the same time reducing the cost. Such a reduction in cost would not increase the overall project time, but the savings might help to offset the increased costs of shrinking other duration times.

By means of this sort of approach, overall project times can be reduced, and total costs minimized. In a simple network as discussed here, no particular difficulties will arise, but in larger networks the number of alternatives will be very great indeed, and it may therefore be necessary to employ a computer for this work.

The Danger of the Cost-Slope Concept

The concept of "cost-slope" is appealing in its simplicity. However, it must be pointed out that—

   (*a*) It is frequently extremely difficult to obtain reliable figures for the changes in cost resulting from changes in duration time;

(*b*) The relationship between cost and time is not a simple one. Multiplying labour time by wage cost is obviously inaccurate and, moreover, to "extend" the resultant labour cost by a constant overhead factor can be equally misleading, since the reduction in time may be obtained, for example, by hiring special plant which has a non-linear hiring rate.

These difficulties make it dangerous to try to deduce general *Time-Cost* curves, or, to put it another way, to assume that cost slopes are constant. For *short* time intervals this assumption can be reasonable, but it is desirable to examine it very closely. All this, of course, is true whether CPA is used or not, but employing CPA has the great advantage that investigations can be directed to the critical activities.

The Relationship between Time and Labour Employed

Duration times cannot be reduced indefinitely by increasing resources. For example, in digging a hole it may well be that two men can carry out the work in less than half the time that one man can carry it out, since work can be efficiently divided. However, three men may not show the same reduction in performance time, and a fourth man may well slow up the work since his physical presence may impede the other workers.

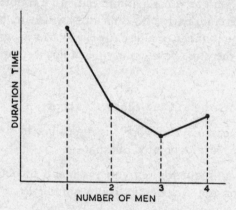

There is thus a minimum time below which it is not possible to reduce the duration time of an activity.

The Final Network

The final network, after a reduction process has been carried out, may well be considerably different from the initial network. The logic may have changed, duration times altered and new critical paths created. Illogicalities may have been introduced, and it is worth retesting the network by checking every activity once more against the two questions—

    (i) What had to be done before this?
    (ii) What can be done now?

Check especially any "cross-road"—

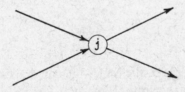

situations to ensure that *all* emerging activities do depend upon *all* entering activities.

QUESTIONS TO BE ASKED WHEN REDUCING THE PROJECT
    TIME

                    PURPOSE?
                    PLACE?
                    SEQUENCE?
                    PERSON?
                    MEANS?
                    RISK?
                    "TRADE-OFF"?
                    COST?

CHAPTER NINE

THE ARROW DIAGRAM AND THE GANTT CHART

A WELL-TRIED and very useful way of representing work which must be carried out is by the bar or Gantt chart. For those readers who are not familiar with this type of chart, a brief description is given here. This treatment does not pretend to be anything more than introductory, and those wishing to study the subject further are recommended to read Wallace Clark's book *The Gantt Chart*.

The Gantt Chart

In this type of chart, the time which an activity should take is represented by a horizontal line, the length of the line being proportional to the duration time of the activity. In order that

TIME / ACTIVITY	WEEK NUMBER																	
	1	2	3	4	5	6	7	8	9	10	11	12	13	14	15	16	17	18

several activities can be represented on the same chart, a framework or ruling is set up, giving time flowing from left to right, the activities being listed from top to bottom (as shown above).

Assume for the sake of simplicity that there are three activities, *A*, *B*, and *C*, which must be carried out in sequence, and that the duration times are—

Activity *A* . . . 4 weeks
,, *B* . . . 6 ,,
,, *C* . . . 5 ,,

This would be represented on the Gantt chart as—

TIME ACTIVITY	WEEK NUMBER																			
	1	2	3	4	5	6	7	8	9	10	11	12	13	14	15	16	17	18		
A																				
B																				
C																				

which would show quite clearly how work should progress. Thus, by the end of week 8, the whole of activity *A* and two-thirds of activity *B* should be complete.

To show how work is actually progressing, a bar or line can be drawn within the uprights of the activity symbol, the length of the bar representing the amount of work completed. Thus, if 50 per cent of an activity is complete, then a bar half the length of the activity symbol is drawn—

This gives a very simple and striking representation of work done, particularly if a number of activities are represented on the same chart, as shown opposite.

If this chart has been correctly filled in, and it is viewed at the end of week 7 (denoted by two small arrows at the top and the bottom of the chart), then the following information is readily apparent—

Activity A should be complete and, in fact, is so,
    ,,   B should be 50 per cent complete, but, in fact, is only 17 per cent finished,
    ,,   C should not be started and, in fact, is not started,
    ,,   D should be 62 per cent complete and, in fact, is only 50 per cent finished,

Activity E should be 17 per cent complete and, in fact, is 50 per cent finished,
    ,,   F should be complete and, in fact, is not started,
    ,,   G should be 87 per cent complete and, in fact, is complete,

or, briefly, incomplete bars to the *left* of the cursor mean under-fulfilment, whilst those to the *right* mean over-fulfilment. By the use of codes and/or symbols, the reasons for any delays can be displayed, and the whole chart can be very succinctly informative, combining both planning and recording of the progress function. For many tasks the Gantt chart is unsurpassed, and its use has been very highly developed.

The Difficulty with the Gantt Chart

The problem that arises with which the Gantt chart cannot easily deal is that of inter-related activities. Whilst the chart above shows seven discrete activities $(A, B, C \ldots G)$, it does not easily show any inter-relationships between the activities. It is possible in small-scale work to "tie up" bars by dotted horizontal lines, but if more than a very few activities are concerned then the chart becomes so muddled as to be useless.

As an example, consider a project which is only complete when three activities C, H and I are complete. H cannot start until an activity D is complete, and I cannot start until activities E, F and G are complete. G cannot start till D is complete, which in turn cannot start until activity B is complete. Activity E must follow activity B, and activity F must follow another activity, A. The whole project starts with activities A, B and C being started. The duration times of the various activities are—

Activity	Duration (Weeks)
A	16
B	20
C	30
D	15
E	10
F	15
G	3
H	16
I	12

Attempting to represent this on a Gantt chart could result in a diagram as shown on page 74 (upper). This does not indicate, for example, that activity I *necessarily* depends on activities E, F and G, or that the whole project must wait upon the completion of activity C.

It may be possible to redraw the chart to show some of these inter-relationships, but within the Gantt framework it is not possible to show them all.

TIME	WEEK NUMBER									
ACTIVITY	5	10	15	20	25	30	35	40	45	50
A										
B										
C										
D										
E										
F										
G										
H										
I										

Inter-relationships and the Arrow Diagram

As we have seen, the arrow diagram abandons length as a measure of time and concentrates on the logical relationships between activities. To do this, the horizontal base is also lost, and the result of the project represented above by the Gantt chart is the arrow diagram which has been so exhaustively discussed—

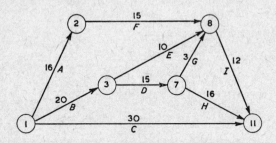

This shows very clearly the necessary inter-relationships, and its analysis can provide much valuable information. It does, however, lack the great assets of the Gantt chart—

(1) that length represents time; and

(2) that the chart *itself* can be used to record the progress of work.

Reconciling the Gantt Chart and the Arrow Diagram

Since the bar-chart representation has such considerable advantages, it is desirable to be able to convert the arrow diagram into a bar chart. Essentially this is done by using the head and tail numbers to show the logical linkages between activities. To illustrate the technique, the above example (which has been extensively used already) will be redrawn as a Gantt chart. Readers who have any difficulty in understanding float will find the translation into the Gantt chart particularly valuable.

Step One

List activities in order of increasing head numbers. Where two or more activities have the same head number, arrange these in order of increasing tail numbers. This, for our example, will give a list as follows—

Activity	Duration
1— 2	16
1— 3	20
3— 7	15
2— 8	15
3— 8	10
7— 8	3
1—11	30
7—11	16
8—11	12

Step Two

Construct a Gantt chart framework with time scale along the top, head numbers down the left-hand side.

Step Three

Set off the first activity in the above list, putting its left-hand

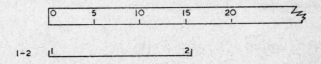

end on the 0 weeks column. Mark the tail and head numbers at the beginning and end of the bar.

Step Four

Set off the second activity on the list, aligning the tail number with the head number of the previous activity, *providing* these numbers are the same. If they are not, then align the tail number with the preceding tail number with which it coincides—

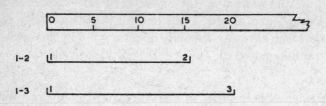

Step Five

Repeat steps three and four for all the activities in turn, aligning tail numbers with previous head numbers where possible, alternatively tail numbers with previous tail numbers. This will then give a chart as shown opposite. (*Note:* A simple rule for drawing the chart is "Match the tail number with that same number which is farthest to the right.") Any dummies must be included as single upright lines.

This chart, although in bar form, is equivalent to the original arrow diagram, providing that it is remembered that the bars are linked together when they have a common number, one at the beginning and one at the end of the bar. Thus, bar 3—8 is linked to bar 8—11, so that if 3—8 is displaced to the right, it can only "move" until its end (—8) coincides with the beginning (8—) of bar 8—11. This then enables the critical path and various floats to be determined without any calculations.

Determination of Critical Path

Clearly, the critical path lies between the point farthest to the right (in this case the 11 in activity 7—11) and the point farthest to the left (in this case any 1). Starting with the farthest

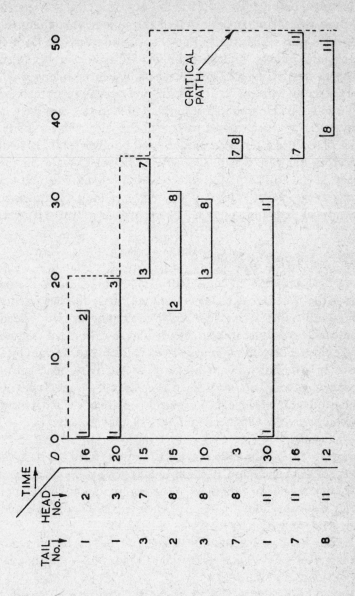

right-hand point, draw a vertical line upwards until it meets the farthest right-hand point carrying the number matching the left-hand number on the first activity—in this case the farthest right-hand number matching the first left-hand number (7) is the 7 in activity 3—7. The critical path then lies along 3—7, and, starting with the 3, draw a line upwards to meet the next right-hand 3 which is farthest to the right—in this case the 3 in activity 1—3. The critical path then lies along this activity. This procedure is repeated until the first number (1) is reached, which in this case is with the activity shown (1—3). In the diagram the critical path is shown with a broken line, and lies along 1—3—7—11. This dotted line is used only to emphasize the critical path and should not be taken as any form of barrier.

Determination of Float

Critical activities, of course, may not "move," but activities not on the critical path can float until they meet another activity whose tail number coincides with the floating activity's head number. It is simpler if these calculations start from the bottom of the chart. For example, *activity* 8—11 can move to the right by 1 week. In so doing it will not meet any other activity, hence it can be said to have a Total float, *and* a Free float of 1 week. It has no Independent float since its movement can affect the movement of other activities with head number —8, that is, activities 2—8, 3—8, 7—8.

Activity 7—11 cannot float at all, since it is "fixed" at its head by the critical path, and at its tail by activity 3—7.

Activity 1—11 can float by 21 weeks and in so doing it will not affect any other activity at all; hence it has Total, Free and Independent float of 21 weeks.

Activity 7—8 can float only if activity 8—11 has floated. Hence, 7—8 has a Total float equal to that of 8—11 (that is, 1 week) and no Free float.

Activity 3—8 can float until it reaches 8—11, that is by 8 weeks, and by a further 1 week (making a total of 9 weeks in all) if

8—11 floats by 1 week. Hence 3—8 has a Total float of 9 weeks and a Free float of 8 weeks. Similarly 2—8 can also be shown to have Total and Free floats of 8 and 7 weeks respectively.

Activities 3—7 and 1—3 cannot float, since they are "fixed" at both ends by the critical path.

Activity 1—2 has float only when other activities float; thus it has no Free float. If, on the other hand, 2—8 and 8—11 float to their fullest possible extent (that is, 8—11 takes up a position between weeks 39 and 51, and 2—8 takes up a position between weeks 24 and 39) then 1—2 can float so that it takes up a position between weeks 8 and 24; thus it has a Total float of 8 weeks, but since it is dependent for this on other activities floating, then it has no Free float.

The results derived above, of course, are identical with those obtained by using the "Rules for Calculating Float." They are set out here at some length, since not only do they show how an arrow diagram is converted to a Gantt chart, but the method in general seems to give a readily comprehensible significance to the various types of float.

Using the above method, it is possible to translate any arrow diagram on to conventional Gantt paper or on to any one of the many proprietary planning boards. Some workers in this field move straight from the arrow diagram to the bar chart and derive all information from this chart. This reduces the calculations—with their accompanying tedium—enormously, but of course it can only be done *once the arrow diagram is drawn*.

The Sequenced Gantt Chart

A useful and economical modification to the normal Gantt chart is that which the author's colleague, R. R. Ritchie, has called the Sequenced Gantt Chart, in which sequences of activities are collected into continuous bars. For example, if we consider the network and Gantt chart already used, in the Sequenced Gantt Chart the critical path is drawn as one continuous bar—

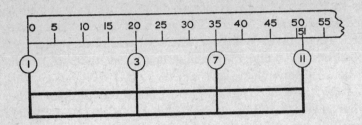

Other sequences are then drawn off the critical path, their starts and finishes being determined either by the events on the critical path or by common events on non-critical sequences. This will give a chart as—

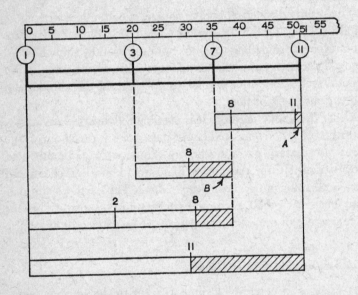

The shaded portions indicate the available float: for example activity 3—8 has a free float of 8 weeks (shaded area B) and a total float of 9 weeks (shaded area B + shaded area A). Similarly, activity 7—8 has no free float since it has no shaded area immediately to its right, but it has a total float of 1 week since it can "shunt" activity 8—11 into shaded area A. The "start" of a sequence is fixed by the earliest start time of the first activity in the sequence, whilst the "finish" of a sequence is fixed by the latest finishing time of the last activity in the sequence. The time between this start and finish is the maximum available time, whilst the sum of the durations of the activities is the necessary time.

The Time-Scaled Network

Yet another method of translating an arrow diagram into a Gantt chart is the time-scaled network, and this is particularly simple if the network is drawn according to the "activities parallel to the edge of the paper" recommendation made on page 16. Thus, if the standard network is re-drawn—

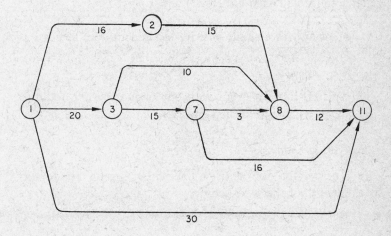

it can be readily drawn to a time scale. However, any network can be time-scaled by the following means—

Start at initial event and draw to scale all opening activities, identifying them by their event numbers. It is desirable at this stage to space these well out on the page.

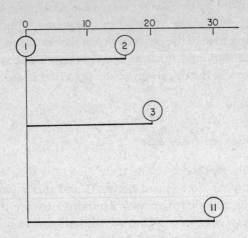

Each activity will then either—

    (*a*) *continue* as a single activity (1—2 continues as 2—8);

    (*b*) *burst* into two or more activities (1—3 bursts into 3—7, 3—8);

    (*c*) *merge* into an event with one or more other activities (1—11 merges into event 11 with 7—11 and 8—11).

Proceed as follows—

(*a*) In the case of "continue" activities, extend the activity by the length of the next activity—

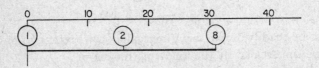

(*b*) In the case of "burst" activities, draw a single vertical line and from this draw to scale the merging activities—

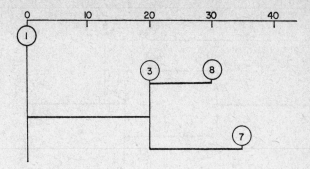

(*c*) In the case of "merge" activities, wait until all activities merging into a common event have been drawn—for example, wait until activities 2—8, 3—8 and 7—8 have been drawn, and then draw a vertical line to form a "barrier" across the end of the activity which extends furthest to the right. Join all the merge activities to that fence by means of dotted lines. These lines represent free float.

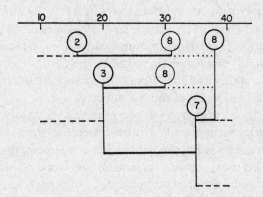

Repeat the above until the last event is reached. Redraw if desired to emphasize special organizational or resource features.

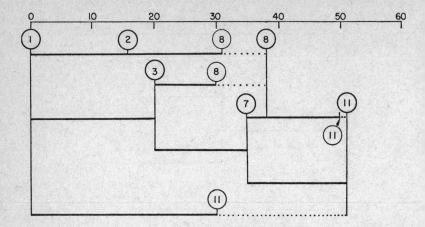

Whilst the modified Gantt charts can be drawn by shifting the bars on the normal Gantt chart into convenient positions, it is often quicker to calculate the event times ($E =$, $L =$) and from these determine the critical path and the boundaries of the non-critical sequences. It is sometimes more useful to position the critical path through the centre of the chart, and some practice is needed if the most appropriate disposition of the various sequences is to be obtained. It is not always necessary to draw the whole chart: parts, by time, function or resource are often useful, and these are simply drawn if it is remembered that the "start" of a sequence is fixed by the $E =$ at the tail event of the first activity and the "finish" of a sequence is fixed by the $L =$ at the head event of the last activity.

The Modified Gantt charts are particularly suitable for use with planning boards, since they use fewer lines than the normal Gantt chart. By writing the description of the activity upon the bar, and by using colours to display different resources and float, a very vivid representation can be obtained, and loading (resource allocation) by inspection can be relatively simply carried out.

In order that any reader who so desires may practise converting arrow diagrams into Gantt charts, the three simple networks

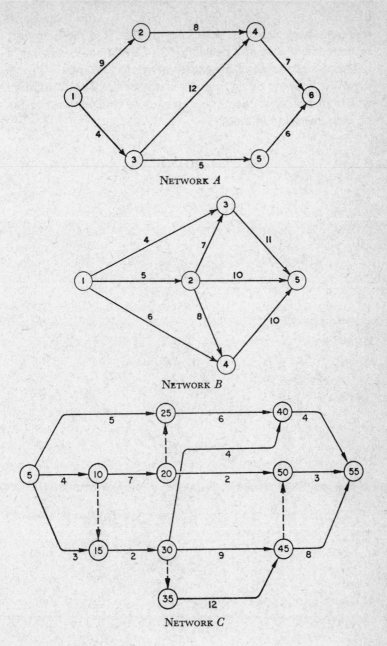

NETWORK A

NETWORK B

NETWORK C

used in Chapter 5 are repeated on page 105, with their Gantt representations shown on pages 197 to 202. It is most strongly recommended that any newcomer to CPA should carry out a full analysis of some of his earlier networks and then translate them into one or other form of Gantt chart. The translation will both act as a check on calculations and emphasize the physical meaning of float.

CHAPTER TEN

LOADING—I BASIC CONSIDERATIONS

IN all the discussions so far, it has been assumed that *time* is the most important factor, and in very many cases this is indeed so. However, it may well be that not only is time important, but the resources employed are equally or more important; for example, the number and skill of the people employed, or the special equipment used may impose severe restrictions.

This problem is one which has been well known to the Production Control engineer since any form of activity has been undertaken. The name given by Production Controllers to this aspect of their work is *Loading*; regrettably, new names have been devised by some of the earlier CPA workers, amongst them *Manpower Smoothing*, and *Resource Allocation*. In the present text *Loading* will be used.

Loading is defined as "the assignment of work to an operator, machine or department," and is a most important feature of producing a time-table. When too much work is required of a work source, the work source is said to be *overloaded* whilst if too little is needed it is said to be *underloaded*. Ideally, the work required should be exactly equivalent to the work available, when the work source is *fully loaded*. This is an ideal situation which is seldom, if ever, encountered except in large-scale flow production, where it is possible to adjust supply and demand in order to reach some form of parity. In the type of work for which CPA is most useful, it is frequently impossible to adjust both supply and demand, and some form of compromise is essential. This usually takes the form of *underloading* since this, at least, produces an acceptable result with respect to time; that is, the promised delivery date can be met. Deliberate *overloading* is foolhardiness to the point of sheer irresponsibility. Projects whose starting and finishing dates are fixed are said to be "time-limited," whilst those where available resources are limited are "resource-limited."

Work Required

The loading function requires, as a point of departure, a statement of the work required. This can only be given in terms of man- (or machine-) hours; it should not be given in any other units *unless* those units are readily and acceptably capable of being translated into resource-hours. For example, it may be that to dig a hole 4 ft × 4 ft × 6 ft in a particular location would take one man twelve hours. The work content should then be specified as 12 man-hours of work, *not* as 96 cubic feet of digging, unless it is well established that—

$$12 \text{ man-hours of digging} = 96 \text{ cubic feet}$$
$$\text{or } 1 \text{ ,, ,, ,, ,, } = 8 \text{ ,, ,,}$$

Similarly, if a designer is committed to design three transformers, it is meaningless to state that his work load is three transformers unless it has previously been established that one transformer is equivalent to so many hours' work.

The work required—that is, in more usual terms the work content—is thus specified in resource-time. More than this, it is necessary also to specify the method employed; the work content of the above 4 ft × 4 ft × 6 ft hole is said to be 12 man-hours when one man is digging. If the method of working is changed, for example by adding another man to clear away the loose earth, the total time taken may be reduced to 5 hours—that is, the work content is then 10 man-hours.

In order to be able to compare various methods, it is often useful to specify that the work is carried out by standard men at standard rates of working. These amiable fictions are primarily useful in planning, and are never found in practice.

A discussion of *standard* in this context will be found in any Work Study text.

Work Available

To complete the task of loading, it is necessary to know the amount of work *available*—that is, the *capacity* available. This, too, must be specified in resource-time, but in order that the completed load will not be fictional it is essential that the capacity should be strictly realistic. Thus, if 100 men are employed, it is unwise to assume that the available capacity each week is 100 man-weeks. In calculating available capacity it is necessary to know—

(*a*) the usual efficiency of working;
(*b*) the anticipated sickness or absenteeism;
(*c*) existing commitments;
(*d*) anticipated maintenance work which is to be done;
(*e*) holidays;
(*f*) any other limitations on working; for example—

(i) confined space;
(ii) limited machine capabilities.

In practice it may well be that some of these factors may be ignored, but it is probably wise to consider each of them as a matter of routine; the number of projects which have not been completed on time because the presence of the summer holidays was forgotten is not on record, but it is uncomfortably high.

Calculation of Load

Once the work available and the work required are known, the ground is clear for the calculation of the loadings on the available men or machines. This is a task which is extremely difficult to carry out with any degree of certainty, since two problems arise.

(1) The Problem of Optimization

It is often loosely stated that an "optimum" load is calculated. The problem here is that it is frequently difficult, if not impossible, to decide which feature should be optimized. Thus, if to complete a task it is necessary to utilize men and machines, and also to purchase material, there are at least four different aspects to consider. Should the programme be constructed to—

      (a) make maximum use of labour;

or (b) make maximum use of machines;

or (c) hold purchased materials for as short a time as possible;

or (d) increase customer goodwill by reducing the overall time as much as possible?

Regrettably these requirements are often in conflict, and, since it is difficult to assign objective values to any of these, and even more difficult to foresee the inter-relationships between the various factors, the decision must often be made on arbitrary grounds.

The essential feature, of course, is that the problem must be considered and a decision must be made. It is sometimes convenient to rank the various factors, and then attempt to optimize "in sequence," i.e. try first to optimize factor 1, then factor 2 and so on, not allowing any succeeding factor to affect any of its predecessors.

(2) The Problem of Alternatives

The problem here is one which arises from the inter-action of one job upon another. For example, if in a department there are three operations, A, B, and C, to be carried out, these operations being independent of each other, then it may be seen that there are six possible sequences in which work can proceed—

(1) Do *A* first, followed by *B*, then by *C*
(2) ,, *A* ,,   ,,   ,,   *C*, ,,   ,, *B*
(3) ,, *B* ,,   ,,   ,,   *C*, ,,   ,, *A*
(4) ,, *B* ,,   ,,   ,,   *A*, ,,   ,, *C*
(5) ,, *C* ,,   ,,   ,,   *B*, ,,   ,, *A*
(6) ,, *C* ,,   ,,   ,,   *A*, ,,   ,, *B*

Similarly, if there are four operations, there are twenty-four possible sequences, and if there are N operations, there are $N!$ possible sequences, and $N!$ increases very rapidly.

Of the possible sequences one (or more) may produce the solution which is "optimum" according to the rules laid down, but to *know* this it is necessary to calculate all the possible sequences. The difficulty of doing this is seen clearly if $N!$ is set out for the first few natural numbers—

N	$N!$
1	1
2	2
3	6
4	24
5	120
6	720
7	5,040
8	40,320
9	362,880
10	3,628,800

and to have ten tasks which should be carried out in one department/section during any one planning period is extremely modest. Further, only one department has been considered. If there are a number of departments, then the number of possible sequences becomes astronomical.

CPA is of some value here, since it is possible to concentrate upon the critical activities, and this may well reduce the choice very considerably. Even so, it may not be possible to examine all solutions, and it is often better to take an *acceptable* solution—that is, one which is workable and does not offend any optimization rules too much—than to spend a great deal of effort in trying to find the probably unrecognizable optimum solution.

Whilst it is probably impossible to lay down any fundamental laws on loading, it would seem reasonable to—

(1) define the optimization rules, and then

(2) use an "acceptable" solution,

in order to reduce the loading problem to easily manageable proportions. It should be realized that, except in limited cases, even a computer cannot calculate all the possible combinations. A manual method is described in the next chapter.

A Limited Case

The general problem of loading is, as we have seen, an extremely complex one. Solutions can be found in some special cases, and these will often depend upon the use of fairly sophisticated mathematics, for example linear or dynamic programming. For much work, however, no ready solutions are available, and so it is necessary to use purely empirical methods.

The Load as a Histogram

It is convenient to represent the load as a histogram—that is, a vertical bar graph, the length of the bar being proportional to the load. For example, if the weekly load in a department which has a capacity of ten man-weeks is—

Week No.	Load (man-weeks)
1	6
2	7
3	8
4	10
5	12
6	6

this would be represented by—

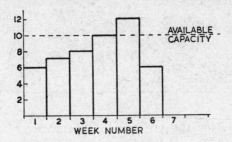

which shows very clearly that the department is underloaded in weeks 1, 2, 3 and 6, fully loaded in week 4 and overloaded in week 5. (*Note.* Some workers in this field refer to the available capacity as the manpower ceiling, so that week 5 exceeds the manpower ceiling.) Although this representation gives no more information than the corresponding sets of figures, it has the usual virtue of a graphical representation, namely great vividness, and in practice it is found that it is almost invariably easier to work in histograms than in numbers, even though the histograms have been derived from the numbers. With experience, a great facility is obtained in viewing a histogram and assessing whether a "peak" (i.e. an overload) can be toppled into a "valley" (i.e. an underload) in order to "smooth out" the loading.

Drawing the Histogram

The simplest way of drawing a histogram is probably found by drawing the appropriate Gantt chart and, by running down each time division, adding up the usage of the various resources. For example, consider the network which has been so frequently discussed—

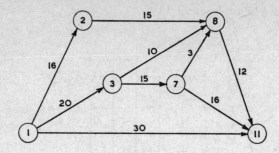

and assume, for the purposes of illustration, that the only resource used is *men*, and that each activity requires manpower as follows—

Activity	Duration	Men
1— 2	16	2
1— 3	20	6
1—11	30	4
2— 8	15	3
3— 7	15	2
3— 8	10	5
7— 8	3	2
7—11	16	4
8—11	12	4

Redraw the Gantt chart, as described in an earlier chapter, inserting the man requirement on each activity bar within a circle. Then, by running down each week, it is possible to add up the manpower requirements very simply, and these can then be plotted on a histogram which, for the sake of convenience, will be shown beneath the Gantt chart (opposite).

In practice it is possible, by the exercise of a little common sense, to reduce the number of additions; for example, it is clear that the loading for weeks 1 – 16 is the same, so that one addition (2 + 6 + 4) suffices.

If the capacity is inserted also on the histogram, the labour situation is very clearly shown. Assume that the available capacity is 10 man-weeks, and that all men are interchangeable. The dotted line shows this capacity and the over- and under-loading.

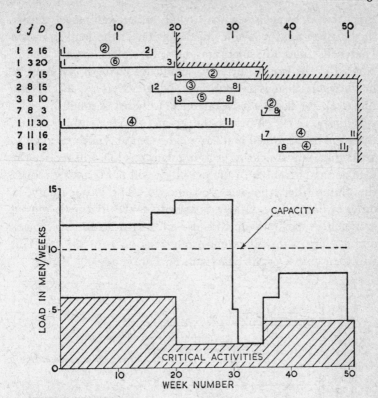

Smoothing the Load

The situation revealed by the histogram is one which is completely unacceptable. For 30 weeks the load exceeds the
capacity, which can have only one result, namely that activities
will take longer than planned, and the overall project time will
increase. For 21 weeks the capacity exceeds the loading, and this
will mean that men are idle. Clearly it is desirable to try to shift
some of the earlier over-load into the later under-load. If this
could be completely done, then the load would be "smoothed."

This problem is an extremely familiar one to all those who have
been in charge of the organization of the disposition of labour.
Virtually intractable, CPA does assist by providing guide-lines
along which to work. Of the various activities some are fixed in

time (that is, some are critical), whilst others can move (that is, they possess float). The critical portion of the load is shown "hatched," and any smoothing must take place in the "plain" portion of the load. Further: significant changes can only be made where float is substantial. Thus, activity 7—8 has only one week free float; its overall effect is therefore small.

Activity 1—11 possesses the greatest float, and it should therefore be examined first. It will be seen that its duration is so great that, whilst floating it as much as possible will reduce the load at the beginning of the period, it will not seriously reduce the "lump" between weeks 20 and 30. The next activity, in order of magnitude of float, is activity 3—8. If this is moved as much as possible, activity 8—11 will advance by 1 week and activity 3—8 will extend from week 29 to week 39. This will then give a Gantt chart and histogram as—

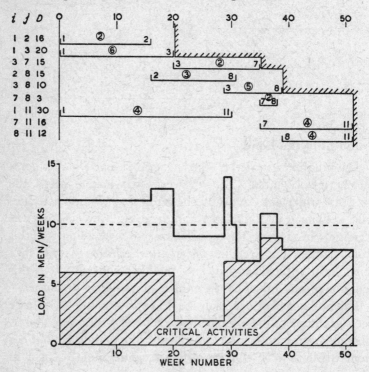

The overload has been completely removed during weeks 20 – 29, and a small overload has been introduced during weeks 35 – 38. All float has been removed from activities 3—8 and 8—11. The remaining activity with any substantial float is activity 2—8. Shifting this forward by four weeks would reduce the "lump" during weeks 16 – 20 and fill the "trough" during weeks 31 – 35. The chart and histogram would then look as follows—

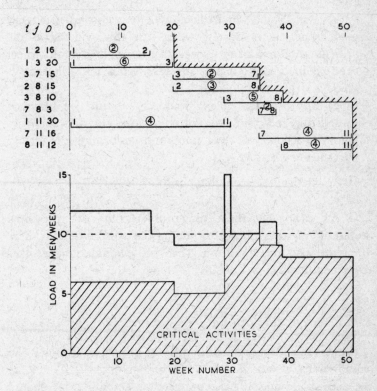

Moving any other activities—and there are only activities 1—2, 1—11 and 7—8 which *can* be moved—would produce no significant change in the load. Hence, the above arrangement is that which gives least overload and, hence, using this as a criterion, is the "best" arrangement.

5—(B.818)

(Note. In practice, the virtue of any particular arrangement can only be judged within the context of the local circumstances.)

The above discussion is offered as an illustration of the kind of thinking which lies behind a loading task. The answer arrived at is not ideal, but answers seldom are in practice. The result, however, does give a sound basis for further consideration by which the problem can only be resolved managerially. For example—

(1) The "spike" in week 29 can be removed if activity 3—8 is advanced by one week, but this will move the critical path forward by 1 week and hence the overall project time will increase from 51 to 52 weeks. Is this desirable/acceptable?

(2) The overload during weeks 1 – 16 can be removed by splitting activity 1—11 into two parts, the first 20 weeks long, the second 10 weeks long, and performing the second part during weeks 42 – 51.

Is this desirable/acceptable/possible?

CPA does not solve the loading problem, but it does provide a method for systematically examining the possibilities. If there are more resources than one, then the examination becomes correspondingly more difficult.

The Effect of Smoothing

What is the effect of smoothing on the project as a whole? To examine this, assume that the last situation above is taken to be the acceptable one. This requires that activity 3—8 should not start until week 29, and that it must finish on week 39. Thus, the earliest and latest start time for activity 3—8 is week 29, and the earliest and latest finish time is week 39 and, since the duration time is ten weeks, all float has disappeared. Effectively, another activity, "wait for availability of labour for activity," has been inserted, and the relevant part of the arrow diagram has changed from—

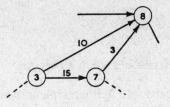

to—

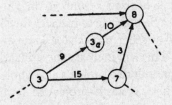

and a second critical path

1—3—8—11

created. Similarly, fixing the starting time of activity 2—8 at not earlier than week 20 has removed the float in the activity. The effect of this sort of action is to reduce the freedom in the network as a whole, while improving the utilization of labour.

Scheduling

Smoothing enables actual dates to be affixed to activities, and this is sometimes known as "scheduling." Thus, whilst initially there were bands of time during which work could start and finish, more starting dates can be fixed as follows—

Act.	Dur.	Start Time Early	Start Time Late	Finish Time Early	Finish Time Late	Float T	Float F	Float I	No. of Men
1— 2	16	0	4	16	20	4	4	4	2
1— 3	20	0	0	20	22	0	0	0	6
1—11	30	0	21	30	51	21	21	21	4
2— 8	15	20	20	35	35	0	0	0	3
3— 7	15	20	20	35	35	0	0	0	2
3— 8	10	29	29	39	39	0	0	0	5
7— 8	3	35	36	38	39	1	1	1	2
7—11	16	35	35	51	51	0	0	0	4
8—11	12	39	39	51	51	0	0	0	4

On comparing this to the original network analysis (page 70) it will be seen that much of the original element of float has disappeared.

Sharing Resources between Projects

Float, as has been stated, is equivalent to an under-utilization of resources. To increase utilization it is common practice to employ the under-utilized resources on another project. For example, it might be possible to transfer the four men needed for activity 1—11 to another project for some or all of the twenty-one days' float. This will increase the utilization of the four men and effectively smooth two projects as a whole. It will, however, have the same effect as single-project smoothing, namely that it will introduce another activity, "waiting for labour from project . . . . .," and in turn this will reduce flexibility.

As with loading, the sharing of resources is a difficult task, and it can only be systematically carried out in very restricted cases. In general the loading and/or sharing of labour is an empirical process, where a workable answer must be accepted, even though it cannot be demonstrated to be the optimum answer. It is in this area that the computer can probably be of maximum assistance, since the computer can work out a large number of solutions very rapidly. It is also this area (which is very similar to areas in other, closely allied, fields) which is receiving enormous attention from research workers, and it may well be that before long this whole problem can be dealt with simply and systematically, although there will always be the need for management to decide on the criteria against which solutions will be judged—a difficult task.

A most useful article in the *Journal of Industrial Engineering*, Volume XVII, Number 4, April, 1966, by Edward W. Davies of Yale University entitled *Resource Allocation in Project Network Models—A Survey* summarizes the solution techniques described in the open literature. The author states "the reasons why there are no generalized analytical solutions for the constrained resource case stem from the difficulties encountered in attempting a mathematical formulation of the problem" and he then discusses some of these problems.

CHAPTER ELEVEN

LOADING—II   A MANUAL METHOD

Despite the difficulties discussed in the previous chapter, the need to produce a solution—albeit an imperfect one—to the loading problem is often encountered. It is useful here to remember Bowman's comment (see *Industrial Scheduling* edited by Muth & Thompson, Prentice-Hall Ltd.) that almost any set of sensible decision rules, if applied consistently, is likely to produce a "better" schedule than an *ad hoc* series of decisions applied unsystematically. The present chapter will set up one such set of decision rules.

Decision Rules

When two or more activities require resources in excess of those available, a *conflict* arises which has to be resolved. For example—

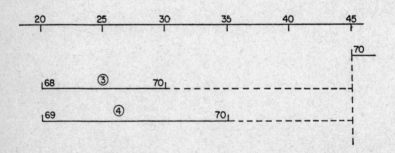

activities 68—70 and 69—70 require 3 and 4 units of a resource respectively. Only 5 units are available, and during the period DAY 20—DAY 30 a total of 7 units are needed. Conflict therefore exists between the two activities and this can be resolved by "awarding" the available resources either to activity 68—70, in which case activity 69—70 must be "slid" to the right, or to activity 69—70, in which case it is the other activity which is moved.

With a multi-activity network it is not possible to test all feasible alternatives, and some routine must be set up to arrive at a decision. This involves the setting up of rules, known as *priority* or *decision* rules, which can rarely, if ever, be justified on any logical grounds, and the results they provide cannot be assumed to be optimal in any way. Many rules can be devised, but for the purposes of illustration a set of rules derived from the work of Dr. Martino will be used here. Whilst these have considerable intuitive appeal, no claim for superiority is made for them.

Manual Resource Allocation

The manual application of a set of decision rules is illustrated here; in cases where it is too difficult or too unwieldy to carry out manipulations by hand, a computer can be used. This requires a *program* which in turn implies a set of decision rules, and the methods employed will be generally similar to the manual method. Considerable insight into computer methods can be obtained by following through the manual technique below.

Step 1. Determine the resources required

For the standard network shown below, assume that the resources required are—

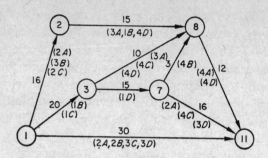

Activity	Duration (days)	Work Content (resource/days)	Type	Resources Required
1— 2	16	32	A	2
	16	48	B	3
	16	32	C	2
1— 3	20	20	B	1
	20	60	C	3
1—11	30	60	A	2
	30	60	B	2
	30	90	C	3
	30	90	D	3
2— 8	15	45	A	3
	15	15	B	1
	15	60	D	4
3— 7	15	15	D	1
3— 8	10	30	A	3
	10	40	C	4
	10	40	D	4
7— 8	3	12	B	4
7—11	16	32	A	2
	16	64	C	4
	16	48	D	3
8—11	12	48	A	4
	12	48	D	4

Note. "Duration," "work content" and "resources required" are all interdependent, any one being deducible from the other two. Circumstances can arise when a different duration must be used in order to change the resource requirement.

Step 2. *Determine the resource ceiling*

In some situations the resource ceiling is known (". . . the maximum number of type A resources which can be made available is . . .") but in other circumstances the requirement is to try to use the minimum possible resources. In this case, proceed as follows—

(*a*) Carry out a forward and backward pass to determine the total project time. If this is acceptable, then loading may proceed. If it is unacceptable, then it will be necessary to reduce the project time by methods similar to those suggested in Chapter 8. DO NOT, AT THIS STAGE, ATTEMPT TO DECIDE ON THE DISPOSITION OF LIMITED RESOURCES. In the above diagram, project time = 51 days.

(*b*) Calculate the total work content for each resource type—

Activity	Resource Type			
	A	B	C	D
1— 2	32	48	32	0
1— 3	0	20	60	0
1—11	60	60	90	90
2— 8	45	15	0	60
3— 8	30	0	40	40
7— 8	0	12	0	0
7—11	32	0	64	48
8—11	48	0	0	48
	247	155	286	286

(*c*) Divide the total work content by the total project time (here, 51 days)—

Resource			
A	B	C	D
4·8	3·1	5·6	5·6

(*d*) Take as a resource ceiling the next whole number above the figure obtained in (*c*) unless that number be itself a whole number, when this is taken as the resource ceiling—

Ceiling for Resource Type

A	B	C	D
5	4	6	6

Step 3. *Prepare a bar chart*

Translate the network into a bar chart with all activities starting as early as possible. The sequenced Gantt chart shown opposite is convenient for this purpose, but any type of chart may be used. Inscribe on the bar chart a statement of the resources required for each activity.

Step 4. *Determine total resource requirements (aggregate resources)*

For each period of time, sum up the resources required and the resource ceilings. Whilst a histogram is vivid, for a number of resources numerical statements are more convenient. This is the process defined in BS 4335 : 1968 as Resource Aggregation— "The summation of the requirements of each resource for each time period, calculated according to a common decision rule."

Note. It is helpful at this stage to rank the resources in importance and to set them down in decreasing order. In the above chart, it has been assumed that resource *A* is of greater importance than resource *B*, which is more important than *C* which in turn is more important than *D*.

Step 5. *Establish decision rules*

Where there is a conflict for resources, award resources according to the following rules devised by Martino—

First Priority .	. in Order of Float
Second Priority .	. in Order of Work Content
Third Priority .	. in Order of Size of Resource
Fourth Priority .	. in Order of Priority of Resource
Fifth Priority .	. in Order of Latest Finish
Sixth Priority .	. in Order of *j* Number

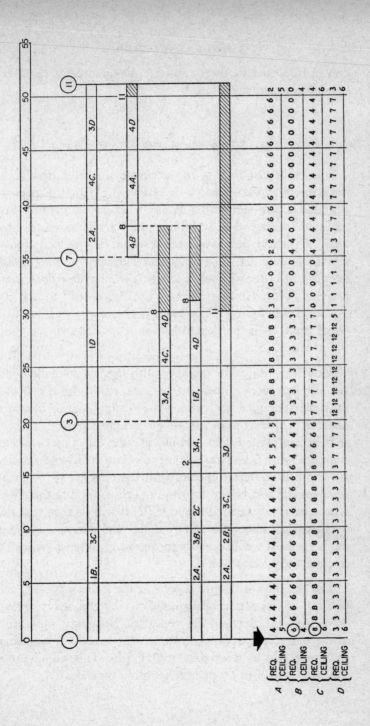

Sets of rules such as these can often be adopted as standard rules within an organization.

Step 6. *Utilize float to cause demand to coincide with capacity*

(*a*) Starting at time 0, run a cursor along the date scale until an overload situation is observed. With the present example this in fact occurs at day 1 where 6 units of *B* are required (ceiling = 4) and 8 units of *C* are required (ceiling = 6). To relieve these overloads it would be possible to move either of activities 1—2 or 1—11. Priority is determined (step 5) in order of float: activity 1—2 has less float than activity 1—11. Hence, resources are "awarded" to activity 1—2 and activity 1—11 is moved. If 1—2 and 1—11 had had equal float, then the next rule would be invoked and then the next and so on.

Move the "movable" activity until its start coincides with the nearest event shown on the bar chart. In the above example there are *two* overloads (*B* and *C*) and these are dealt with in accordance with the resource priority rules— that is, deal with *B* first and observe the effect on *C*, rather than deal with *C* and observe the effect on *B*. Sliding activity 1—11 until its start coincides with the next event (event 2) removes the *B* overload for days 0—16. Checking the effect on the other overloaded resource (*C*) it will be seen that the overload has also been removed for this period. Had this not been so, it would have been necessary to repeat this step to clear the overload on *C*.

The situation is now as shown in the chart opposite, the position of the cursor being indicated by the heavy arrow. Note that the demand for resources is being "squeezed" to the end of the network, in much the same way as pastry is "squeezed" in front of a rolling pin. This piling up is characteristic of this type of scheduling procedure.

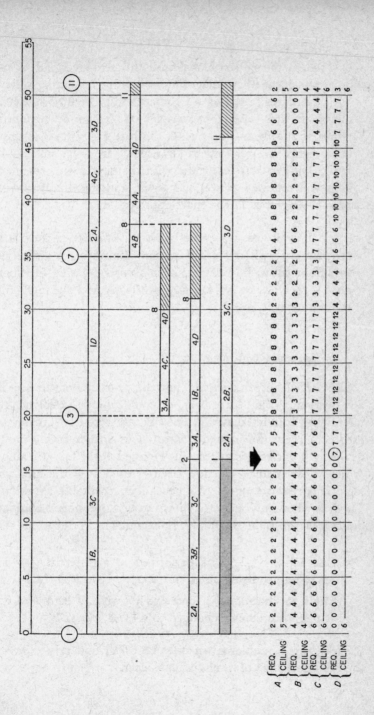

(*b*) Run the cursor along until the next overload is indicated—in the above diagram this is at the beginning of day 17 (*D* requires 7, ceiling 6). Repeat the procedure outlined above. In this case, the overload on *D* can be removed either by sliding 1—11 or 2—8. Of these 1—11 has the least float, hence the resource is "awarded" to 1—11 and 2—8 is slid until its start coincides with the next event, event 3. This will relieve the overload on *D*, and the situation will be represented as in the chart opposite.

(*c*) This process is repeated, and the situation in the chart on page 132 will result. This now represents a common situation: all useful float has been absorbed (using the float on 1—11 will not reduce any overload) and the load still exceeds capacity.

Step 7. *Re-examine basic logic*

The initial network should now be re-examined to see if any overloads could have been, or could now be, resolved by modifying the initial network. The logic and resource statements used so far are the immediately obvious and desirable ones, in the light of circumstances which existed at the first drawing. A new circumstance has now been created, and the original logic and resources used produce an unacceptable situation. Thus, as when any budget is shown to be impossible the initial policies of the plan should be re-examined. For example—

Can activity 1—11 be performed in a different way in order to reduce its demand on resources type *B* and *C*?

Could its duration be increased to, say, 40 days and its resource requirements changed to 2*A*, 1*B*, 1*C*, 1*D*?

Clearly, these questions can only be asked and answered in the full knowledge of local circumstances.

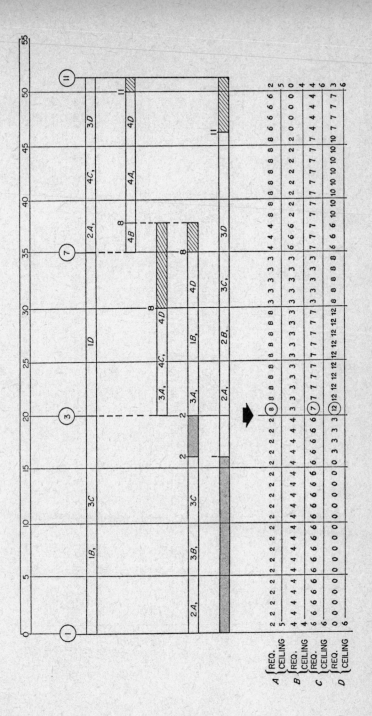

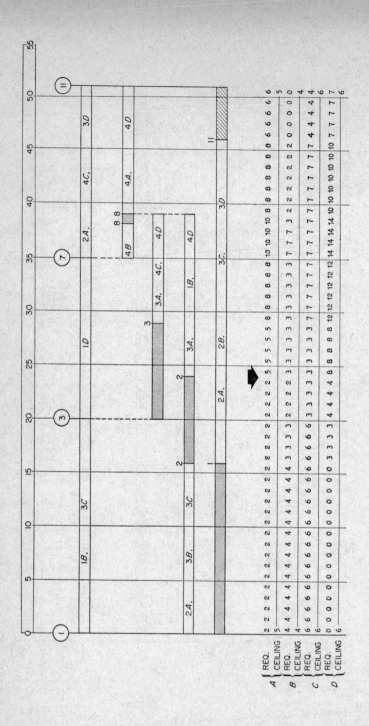

Step 8. *Decide on limitations*

Further action now depends on the taking of a fundamental decision: is the task "time-limited" (that is, must the total project time of 51 days be considered as fixed) or is it "resource-limited" (that is, must the load be kept equal to, or less than, the ceiling)? If the task is time-limited, then new ceilings must be set and the whole business repeated, the increasing of the ceilings being by steps as small as sensible (for example, try $A = 6$, $B = 6$, $C = 8$, $D = 12$). On the other hand, if the task is resource-limited, then the finishing date must be allowed to "settle" in a position determined by the movement of the activities. This will "open up" float not previously available, and will probably change the critical path. (This is illustrated in the diagram overleaf where activity 8—11 is moved to finish at time 56, and float is opened up between time 51 and 56 for activities 7—11 and 1—11 and between time 39 and 44 for activity 3—8). This process continues, again using the rules set down, until no resource ceiling is exceeded.

RESOURCE ALLOCATION PROCEDURE

1. Draw initial network and determine resources required.
2. Determine resource ceiling.
3. Prepare bar chart in "all earliest start" position. Inscribe resources required on network.
4. Determine total resource requirements.
5. Establish decision rules.
6. Utilize float to apportion resources.
7. If result unacceptable, re-examine basic logic.
8. Re-examine time/resource limitations.

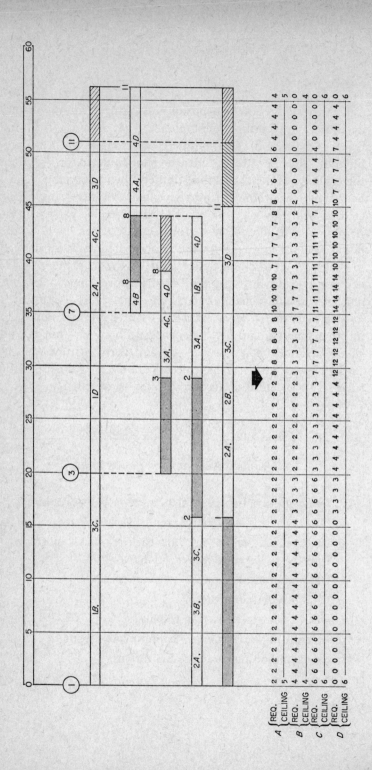

CHAPTER TWELVE

CONTROL AND CRITICAL PATH ANALYSIS

CONTROL is here used in a special sense; it is *not* used to mean supervision or direction, but it *is* used to mean the comparing of that which takes place with that which has been planned to take place. As such, it has the significance given to Control in budgetary control, where the actual expenditure is compared with the planned expenditure, and any differences (the accountant's "variances") are reported to the person responsible for the expenditure.

The Essential Features of Control

An industrial control system appears to have six essential features—

(1) a plan must be made;

(2) this plan must be published;

(3) once working, the activity being controlled must be measured;

(4) the measurements must then be compared with the plan;

(5) any deviations must be reported to the appropriate person;

(6) a forecast of the results of any deviations must then be made, and corrective actions taken to cause the activity to continue in a way which will produce the originally desired result, or, if this is not possible, a new plan must be made.

The above six features appear to be general to any industrial control system, and they should therefore be considered when using CPA as a control technique.

Planning and Publishing

Looking at CPA it will be seen that inherently it contains the first two basic features, planning and publishing, most adequately. In many tasks it is the only possible planning technique available, and its use as a means of communication has already been commented upon; it is an excellent means of publishing a plan.

Measuring

Measuring activity is not an integral part of CPA, although the discipline imposed by constructing the network, and the consequent depth of insight into the project, will indicate how best these measurements can be made. There are, again, a number of general features of control measurements which have emerged from other control situations which appear to apply when using CPA in a control situation.

GENERAL FEATURES OF CONTROL MEASUREMENTS

1. The measurement should be appropriately precise. Any measurement can be increased in precision by an increase in the cost of making the measurement. CPA indicates very clearly which activities need to be precisely measured (those on the critical path) and those which do not need such a high precision. For example, in the network already discussed, measuring activity 1—3 to the nearest day could well be useful, since 1—3 is on the critical path and any "slip" would result in an increase in overall time for the project. On the other hand, activity 1—11 has a total float of 21 weeks, and to monitor this to the nearest day would be uselessly expensive.

2. The measurements should be pertinent. This is quite self-evident, yet the files of industry bulge with data which have been collected and which are not used. It is essential to question the use which will or can be made of any data.

3. The speed of collection of the information must be rapid compared with the time-cycle of the system as a whole. In a project lasting two years, collecting information and processing it every two weeks is probably adequate, since it will allow corrective action to be taken. No general rule can be laid down here, but it must be remembered that, as a project progresses, the time remaining for completion diminishes and, hence, the speed of collection may need to increase. Thus it is quite usual, at the outset of a long project, to receive reports every month but, as time advances, to reduce the reporting time to once a fortnight, later to once a week and eventually to once a day. The cost of collecting information is high.

4. Measurements need to be accurate or of consistent inaccuracy. As with the degree of precision, so with accuracy; accuracy can be bought with increased cost. It is frequently cheaper to accept a measuring technique which is known to be inaccurate but consistent than to attempt to obtain a very high accuracy. Consistent inaccuracies can be allowed for; high accuracy inevitably results in increased cost. Here again, CPA indicates where a high accuracy measurement should be made and where a low accuracy one is tolerable. Thus, activity 1—3 is measured to the nearest day, and ten per cent accuracy may well be useful, whereas activity 1—11, which is measured to the nearest week, could probably tolerate a very much greater error—say fifty per cent.

(*Note.* Although accuracy and precision are related, they are quite separate concepts, and they should not be confused.)

5. The number of data-processing points should be kept as small as possible. Once a measurement is made, it should be passed through as few processing departments as possible. Not only will handling delay the using of the information, but it will inevitably cause distortions which are very difficult to eliminate.

Comparing and Reporting

Some of the ways in which comparing and reporting may be carried out with CPA are noted below. Whilst it is possible to devise many other methods, it is sensible to avoid letting the ingenuity of the method become an end in itself. The author has observed many clever techniques in which the mechanics of the method have obscured the results. *The simplest method is always the best.*

1. THE NETWORK ITSELF

As work progresses, the arrow itself can be marked in some way to indicate that work has been done, although this tends to be unsatisfactory since the arrow length is not related to the duration time. Another method is to use the quadrated node symbol, and insert the "arrival time" in the top segment. Thus, if activities 2—8, 3—8 and 7—8 are all completed by time 39, 39 is inserted in the circle—

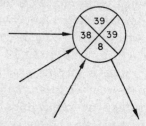

Providing the inserted time is equal to or less than the right hand time, then the project as a whole can be achieved without replanning.

2. THE BAR CHART

If the arrow diagram has been translated into a bar chart, performance can be signalled by drawing a progress bar (see pages 91 and 92). This is both simple and graphic, and in many cases is the most appropriate way of displaying progress information.

3. THE LATEST TIME CHART

It is sometimes felt that the Gantt chart does not adequately show when jobs *must* be finished. Non-critical jobs can "slip" to the critical line without upsetting the overall performance time, but it requires some thought to deduce this from the Gantt chart. To get over this, some workers draw a "latest-time" chart. On this, the events are shown at their latest times on a linear time grid—

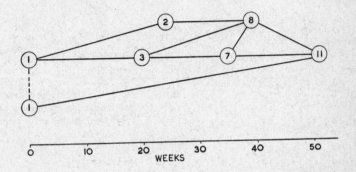

The critical path here appears as a central straight line with the events marked appropriately upon it. The non-critical events are arranged above or below this line, as convenient. Event 2 has a latest time of 24 weeks, and an event circle is drawn at 24 weeks, and circles 1 and 2 are joined to show the relationship between 1 and 2. Similarly all other events are shown at their latest times. Progress can then be signalled by indicating the point which, in fact, any activity has reached. This can be done in a number of ways: for example, by using large, flat-headed pins upon which the event number can be written, so that the difference between the pin and the chart circle will indicate how much more time can elapse. Again, by the use of some ingenuity, all lines can be turned into horizontal lines, linkages being shown by dotted lines.

4. RE-ANALYSIS

In complex projects it is virtually impossible to represent the situation graphically. Under these circumstances, CPA can prove to be of inestimable value. By taking the original network and inserting into it the *actual* times, instead of the expected duration times, it is simple to re-analyse the network and see the effects of the actual work. These effects will be shown up most clearly by a change in float. For example, assume that at the end of week 30 the situation with regard to the sample network is as follows—

| | Duration Time | | |
Activity	Expected	Actual	Notes
1— 2	16	18	Complete
1— 3	20	19	Complete
1—11	30		Not started
2— 8	15	20	Partly complete; re-estimate
3— 7	15	15	Partly complete; re-estimate
3— 8	10	15	Partly complete; re-estimate
7— 8	3		Not started
7—11	16		Not started
8—11	12		Not started

Note that activity 1—11 has a latest start time of week 21, and this has already been exceeded. It is therefore necessary to take cognizance of this fact, which is done here by inserting another activity (1—1a), which is a delay activity of duration 30 weeks. Another way of showing this would be by increasing the duration time of activity 1—11 to 60 weeks.

Analysing the network as previously explained will give the table shown on page 141; this shows quite clearly the effect of the various performances, the principal one being that the critical path has now shifted to 1—1a—11, float being opened up in the previous critical path. The project can now be re-examined to see where and how this slipping can be made good.

An alternative technique is to re-draw the network, leaving out those activities which are complete, and then re-analysing in the normal way, substituting the actual date for the date of the first event. This will have the effect of gradually collapsing the network from the left, and in large networks a progressive simplification will result. It will obviate the need for inserting new "delay" activities. In the example shown, there is little advantage in doing this.

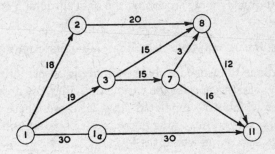

Activity	Duration	Start Time		Finish Time			Float		
		Early	Late	Early	Late	Tot.	Free	Ind.	
1— 2	18	0	10	18	28	10	0	0	
1— 3	19	0	10	19	29	10	0	0	
1— 1a	30	0	0	30	30	0	0	0	
1a—11	30	30	30	60	60	0	0	0	
2— 8	20	18	28	38	48	10	0	0	
3— 7	15	19	29	34	44	10	0	0	
3— 8	15	19	33	34	48	14	4	0	
7— 8	3	34	45	37	48	11	1	0	
7—11	16	34	44	50	60	10	10	0	
8—11	12	38	48	50	60	10	10	0	

5. NEGATIVE FLOAT

If the project is large, and the network complicated, it may be tedious to re-draw. An alternative is to insert the actual (or re-estimated) times, and to put the planned or accepted project duration time as the latest time of the last event. Thus, in the above example, the latest time of event 11 will be taken as week 51 and, hence, the latest finish time of activity 1a—11 will be week 51. Calculations can then be as previously, the result being that activity 1a—11 will appear to have a negative float (–9 weeks). This will indicate immediately that, in order to achieve the total project time of 51 weeks, activity 1a—11 must be reduced in duration time by 9 weeks. In many cases this negative-float technique is most appropriate, since it can be obtained quickly by reprocessing the data through a computer.

6. Time Remaining and the Elapsed Time Arrow

To reduce the work involved in analysis, insert into the network the *time remaining* for the various activities, discarding those activities which are completed (that is, those where the time remaining = 0). Thus, using the example of paragraph 3 above, the times remaining would be—

Activity	Time Remaining
1— 2	0
1— 3	0
1—11	30
2— 8	8
3— 7	4
3— 8	4
7— 8	3
7—11	16
8—11	12

and the network would be—

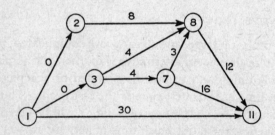

or discarding the zero time activities the network would shrink
to—

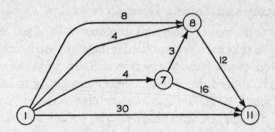

This is a smaller network to handle than the original, but its
float times would be identical to those shown in the table in
paragraph 3. However, the total project time would be shown
to be 30 weeks, and it would not be possible to compare directly
the activity times with the activity times of the original analysis
which were based on a total project time of 51 weeks. This dis-
advantage can be overcome by inserting an "*elapsed time*" arrow
at the beginning of the network, the duration time of this being
equal to the time which has elapsed since the start of the project.
In the example so far used the elapsed time arrow would have a
duration of 30 weeks (since the project is at the end of week 30)
and the network would become—

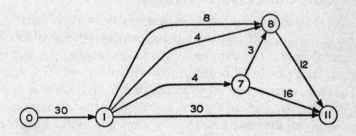

This network, being analysed, would give results directly com-
parable with the original analysis (that on page 70) and any
comparisons required would be easily made. Ideally, of course,

the duration of the elapsed time arrow at the conclusion of the project should be equal to the original project time.

Forecasting and Taking Corrective Action

When performance does not conform with plan, and it is necessary to take corrective action, it must be clearly understood that CPA does not remove any responsibility from the manager concerned; indeed, by causing areas of authority to be clearly distinguished, it reinforces and emphasizes the manager's position. CPA is neither a prophylactic, nor is it a panacea, and any failures to achieve an agreed plan must not be laid at the door of CPA; they will rest, as always, with the manager.

This having been said, it must be pointed out that CPA has a particular use in this field, namely that it will enable predictions of resultant actions to be deduced from present or past action. For example, any "slip" in a critical activity will result in the whole project "slipping." To correct this it may be possible to transfer labour from other non-critical activities, and the consequent effects of this can be clearly seen by considering the arrow diagram, or repassing the modified data through a computer if one is being used. This predictive value of CPA is probably unique amongst planning systems, and certainly is of great value in real-life situations.

CPA and Other Control Systems

All the previous discussions have considered CPA in isolation: this arises from the restrictions imposed by writing a text in general terms. The data used in, and the information derived from, CPA should all be integrated with other control systems. The time taken to perform an activity is of interest not only to the CPA planners but also to the costing department, the wages department, the material-control section, and so on. Failure to co-ordinate the work of various departments will lead to dissipation and duplication of effort, and will also prevent those cross-checkings which can do so much to make up for inadequacies. Not only this—a number of imperfect systems acting in parallel can often produce a total accuracy which one highly perfected system cannot attain, and this can enable economies to be made which do not result in any degrading of information. It is therefore highly desirable that CPA be used *as a control technique* as well as a planning technique, and that its work here should be integrated as closely as possible with all other management controls.

PERT-cost systems

In systems known generically as PERT-cost systems, an estimated cost is obtained for each activity or group of activities. After scheduling—and cash may be one resource considered—a cash outflow is calculated for the life of the project, and subsequent comparison of actual with estimated expenditures then provides a cost control system. The most usual difficulty met with is that costs are often extracted and recorded for elements unlike the network activities, and to ensure that the cost and time elements coincide may require an assault on long-established procedures which are strongly defended.

If the activity/cost estimates are—

Activity	Duration (Weeks)	Cost (£)	Effective Cost/Week (£/Week)
A	16	320	20
B	20	100	5
C	30	660	22
D	15	60	4
E	10	220	22
F	15	150	10
G	3	90	30
H	16	80	5
I	12	360	30
Total		2040	

then a cash outflow "envelope" can be produced, within which the scheduled cash outflow will lie; this and the scheduled outflow are most easily derived from the appropriate bar charts. Below are the "earliest" and "latest" bar-charts, each with the "cost/week" superimposed on the appropriate activities.

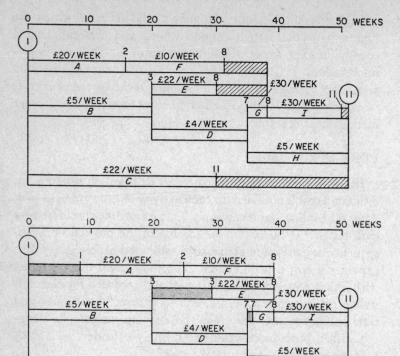

If activities start as *early* as possible, the cash outflow is—

Period (Weeks)	Duration (weeks)	Effective Weekly Rate (£/Week)	Cost for Period (£)	Outflow since Project Start (£)
0—16	16	47	752	752
16—20	4	37	148	900
20—30	10	58	580	1480
30—31	1	14	14	1494
31—35	4	4	16	1510
35—38	3	35	105	1615
38—50	12	35	420	2035
0—51	1	5	5	2040

If activities start as *late* as possible, the cash outflow is—

Period (Weeks)	Duration (Weeks)	Effective Weekly Rate (£/Week)	Cost for Period (£)	Outflow since Project Start (£)
0— 8	8	5	40	40
8—20	12	25	300	340
20—21	1	24	24	364
21—24	3	46	138	502
24—29	5	36	180	682
29—35	6	58	348	1030
35—36	1	59	59	1089
36—39	3	89	267	1356
39—51	12	57	684	2040

These two are results are shown graphically—

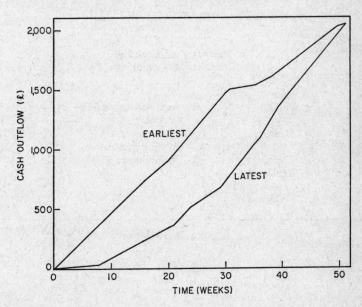

These results, and their summary in the graph represent only the *direct* costs, no allowance being made for any indirect or overhead charges. The representation of these costs can present considerable difficulties. If the overhead rates for each activity or group of activities are known, special "hammock" activities can be inserted to carry the "burden" of the overheads. For example, if activities A and F derive from the same cost-centre, and the overhead rate is £AF/week, then a new activity, activity 1—8, whose duration *is equal to the joint activity-span* of A and F can be inserted and its cost-contribution calculated as above. In practice it is often found that the managerial decisions concerning the apportioning of overheads have either not been made, or that the form in which they have been made is not appropriate for this type of approach. Despite these difficulties, the value of associating time and cost is very great, and on major projects it may well prove invaluable to re-think the whole costing system.

THE FEATURES OF CONTROL
AND CONTROL MEASUREMENTS

1. Plan
2. Publish
   (1) Appropriately precise
   (2) Pertinent
3. Measure
   (3) Fast
   (4) Accurate/consistent
4. Compare
   (5) Minimum handling
5. Report
6. Forecast and Correct

CHAPTER THIRTEEN

ACTIVITY-ON-NODE NETWORKING

THE network diagrams discussed so far require an activity to be shown as an arrow, and the dependency of one activity upon another is given by the relation of the arrows to each other. There is, however, a family of networking systems where the activity is represented by the *node*, and the dependencies by the *arrows*. Of these *Activity-on-Node* systems (AoN) possibly the best known is that called the "Method of Potentials" by Monsieur B. Roy, or simply the "Critical Path Method" by M. J. Fondahl. Method-of-Potentials (MoP) will be discussed in detail here, and brief reference will be made to other systems at the end of this chapter.

Representation of Logic in MoP

The dependency rule in MoP gives configurations as in the following diagrams where it will be seen that MoP (unlike CPA) does not require dummy activities to maintain logical relationships. It is worth referring to the illustration on page 27 where the "all-dummy" method of locating dummies in CPA produces a diagram very similar to the MoP diagram, and indeed, dependency arrows in AoN serve much the same purpose as dummies in CPA.

Representation of Time in MoP

Whilst CPA uses a subscript to the activity arrow to represent the duration time of the activity, MoP uses a subscript to the dependency arrow to denote the dependency time, that is,

SITUATION	CPA	MoP
	Node A	Node B	Node	Node	Dependency	Node
	Activity	Activity	A	B
Activity B depends on Activity A		A → C ← B	A, B → C
Activity C depends on Activities A and B			
Activities C and D depend on Activities A and B			A, B → C, D
Activity C depends on Activity A; Activity D depends on Activities A and B			A → C; A, B → D
Activity K depends on Activity A; Activity L depends on Activities A and B; Activity M depends on Activities A, B and C			A → K; A, B → L; A, B, C → M

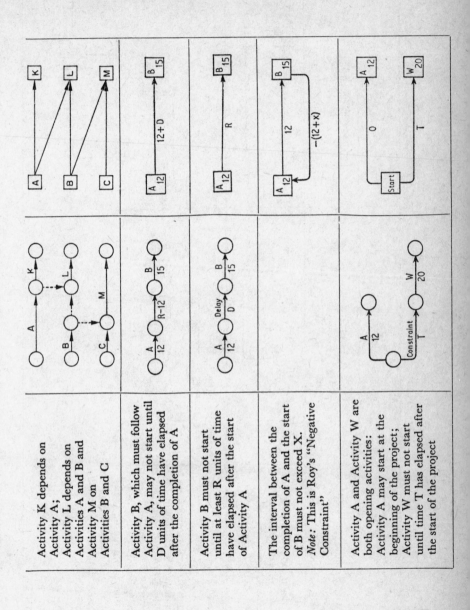

Activity K depends on Activity A;
Activity L depends on Activities A and B and
Activity M on Activities B and C

Activity B, which must follow Activity A, may not start until D units of time have elapsed after the completion of A

Activity B must not start until at least R units of time have elapsed after the start of Activity A

The interval between the completion of A and the start of B must not exceed X.
Note: This is Roy's "Negative Constraint"

Activity A and Activity W are both opening activities:
Activity A may start at the beginning of the project;
Activity W must not start until time T has elapsed after the start of the project

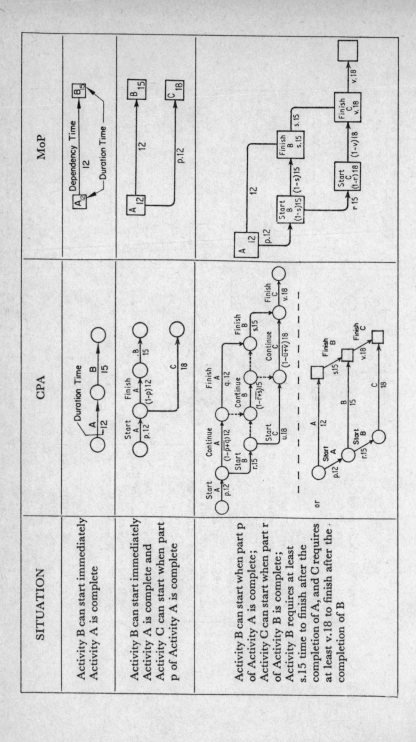

SITUATION	CPA	MoP

Activity B can start immediately Activity A is complete

Activity B can start immediately Activity A is complete and Activity C can start when part p of Activity A is complete

Activity B can start when part p of Activity A is complete; Activity C can start when part r of Activity B is complete; Activity B requires at least s.15 time to finish after the completion of A, and C requires at least v.18 to finish after the completion of B

the time which must elapse between the START of an activity and the START of the succeeding dependent activity. This permits considerable flexibility in showing the time relationsips between activities and constitutes one advantage of MoP over CPA. Since the dependency time is not necessarily the duration time of the tail activity, the duration time may be "lost," and for this reason the author prefers to include the duration time within the activity node. Some typical configurations are given in the diagrams on the facing page.

Milestones in MoP

It is sometimes convenient in the life of a project to identify "milestones" when particular decisions have to be taken or situations reviewed. Milestones usually represent the completion of a number of activities, and in CPA they are represented by events (or nodes). In MoP, events do not exist as such: however, it is possible to represent a milestone in MoP by using a fictitious activity, of duration o.

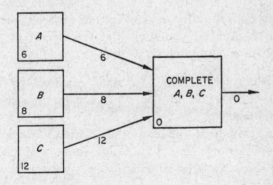

Analysis of an MoP Network

The analysis of an MoP network requires forward and backward passes, identical in manner to those carried out in conventional Activity-on-Arrow CPA. The result of the forward pass is unchanged—it gives the earliest starting time of the activity. The backward pass, however, gives the LATEST STARTING TIME of the activity, *not* the latest finishing time as in A-on-A CPA. The result of a forward and backward pass on the standard network used throughout this volume is shown as follows and should be compared with the diagram on the opposite page.

The meaning of the two node times must be clearly understood: the L.H. node time, calculated on the forward pass, means *the earliest time at which the activity can start*, and the R.H. node time, calculated on the backward pass means *the latest time at which the activity can start*. Thus, node E, from the previous diagram

20	29
E	
10	9

means—

> Activity E, of duration 10, can start as early as time 20 and as late as time 29, assuming that the project as a whole starts at time 0 and finishes at time 51.

Activity E thus has the ability to "slide" or "float" by (29—20) = 9 units of time, and this float figure is entered in the bottom right hand box.

Simple arithmetic enables the earliest finishing time (=earliest starting time + duration time) and the latest finishing time (=latest starting time + duration time) to be calculated—

Activity Description	Time	Start Early	Start Late	Finish Early	Finish Late	Float
E	10	20	29	30	39	9

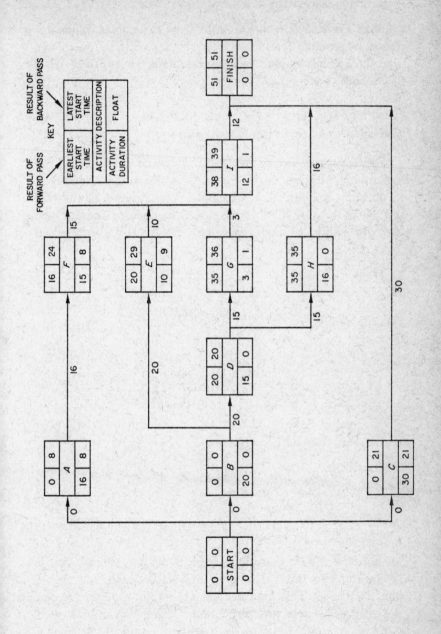

Note. Purists will realize that this is TOTAL float. While it is found in practice that this is adequate for most purposes, free and independent float can be calculated from their basic definitions (page 63).

Calculations Involving Constraints

Positive constraints in calculations, present no difficulty, but negative constraints require care. The situation: "Activity *A* of duration 12 is followed immediately by Activity *B* of duration 15, whilst Activity *K*, of duration 36 is an opening activity with Activity *A* and a closing activity with Activity *B*," is represented by—

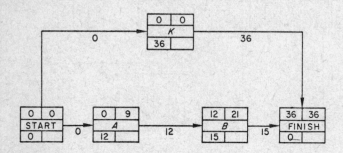

and the analyses for *A* and *B* are—

Activity Description	Time	Start Early	Start Late	Finish Early	Finish Late	Float
A	12	0	9	12	21	9
B	15	12	21	27	36	9

so that it would be possible to finish *A* at time 12 and start *B* at time 21, an interval of 9 time units between finishing *A* and starting *B*. This may not be acceptable: for some reason the interval between finishing *A* and starting *B* must be limited to, say, 4 units of time. A negative constraint of value

$-12 + (-4) = -16$ pointing backwards from B to A would achieve this—

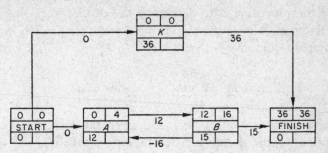

The forward pass is carried out in the usual way ignoring the negative constraint. On the backward pass, Activity B has two emergent arrows (one, the negative constraint arrow). The figure for the latest start of B is determined by—

(a) subtracting the normal (that is, non-negative) dependency time from the latest start time of its head activity

$$36 - 15 = 21$$

and

(b) subtracting the constraint dependency time from the earliest start time of *its* head activity

$$0 - (-16) = 16$$

and then choosing the smaller (16) from these two. The rest of the backward pass proceeds normally, and the resultant analysis for A and B is—

Activity Description	Time	Start Early	Start Late	Finish Early	Finish Late	Float
A	12	0	4	12	16	4
B	15	12	16	27	31	4

Thus, even with A finishing as early as possible (12) and B starting as late as possible (16), the "gap" between A and B cannot exceed 4.

In effect this "ties" A and B to their earliest possible positions. It may be that a later position is more appropriate. In this case the procedure is to determine the earliest start of A and the latest start of B *ignoring the negative constraint* which in this example would give—

Earliest start A o: Latest start B 21

A decision has now to be made as to the most convenient position for the A—B pair, which, for the sake of illustration is assumed to be such that the latest starting time for B is 18. Using this as the "fixed point," the earliest starting time for A is now calculated using the two "entering" arrows (one from start, dependency time o, one from B, dependency time 16). As these are entering arrows a forward pass is being performed, so that the earliest starting time of A is the larger of the two—

$$o + o = o$$

or $$18 + (-16) = 2$$

that is the earliest starting time of A is 2. This is used in place of the previously obtained figure of o which is struck out and the earliest start of B ($2 + 12 = 14$) and the latest start of A ($18 - 12 = 6$) are redetermined.

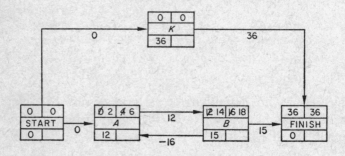

Paired Jobs

There are circumstances when one job must start *immediately* after its predecessor has been completed. In MoP this is represented by—

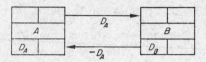

The Critical Path

The critical path for a project, like float, exists whether the project is set down graphically or not. Defining the critical path as that sequence of activities which determines the total time for the project, it can be recognized as that series of activities having minimum float. In the case where the earliest and latest starting time of the final activity are the same, then the critical path is determined by that sequence of activities where the earliest and latest starting times of activities are the same. The most general definition of the Critical Path is, of course, that given on page 61.

Scheduled Times

Scheduled times (". . . a date imposed on a network by circumstances outside the logic of the network") can be inserted in MoP in just the same way as in CPA, and can give rise to the same features—secondary critical paths, negative float—should circumstances dictate.

Gantt Chart Translations of an MoP Diagram

The MoP diagram can be translated into a Gantt chart in a way very similar to the CPA translation. The resulting diagram will, of course, be identical whether MoP or CPA is the originating network.

Resource Allocation

Resource allocation in CPA can be undertaken in one of two ways—

(*a*) by manipulating a derived bar (Gantt) chart;
(*b*) by using an analysis giving the earliest and latest starting and finishing time, and the float, for each activity.

Since either of these can be produced from either CPA or MoP with almost equal facility, the problem of resource allocation is identical in both cases.

Matrix Method of Expressing and Analysing MoP Diagrams

Roy has set down a tabular method of expressing and analysing an MoP diagram which does not require that the diagram itself should be drawn as a collection of arcs and nodes. A matrix can be used for the same purpose, and this has considerable advantages over the tabular form in clarity and simplicity of construction. To illustrate the method the same example as previously discussed will be used, although it must be emphasized that the network need not be drawn as on page 154, the matrix itself can be used to supply all the logical and time dependencies.

Step 1

List all activities, along with their duration times. Add to this list two other activities START and FINISH, each of zero duration times

A—16	D—15	G—3	START—0
B—20	E—10	H—16	FINISH—0
C—30	F—15	I—12	

Step 2

Prepare a square matrix with column headings and row descriptions in accordance with the complete list of Step 1. It is convenient to put START and FINISH at the beginning and end of the rows and columns, but otherwise the orders in which the activities are written down are of no consequence. The column headings represent SUCCESSOR activities and the row descriptions PREDECESSOR activities.

SUCCESSOR

PREDECESSOR		START	A (16)	E (10)	G (3)	H (16)	B (20)	I (12)	D (15)	F (15)	C (30)	FINISH
START												
B	20											
C	30											
H	16											
A	16											
D	15											
F	15											
G	3											
E	10											
I	12											
FINISH												

Step 3

Represent logic by "boxing" the cell at the intersection of dependent activities. Each column except START and each row except FINISH must contain at least one "boxed" cell.

SUCCESSOR

PREDECESSOR		START	A (16)	E (10)	G (3)	H (16)	B (20)	I (12)	D (15)	F (15)	C (30)	FINISH
START												
B	20		□			□	□					
C	30											□
H	16											□
A	16						□					
D	15				□	□						
F	15							□				
G	3							□				
E	10							□				
I	12											□
FINISH												

Step 4

The logic having been set down, dependency times are added by inserting into the "boxes" the time which must elapse between the starts of the predecessor activity and the successor activity. Constraints (either positive or negative) are added just as in conventional MoP. In the example being discussed, all dependency times are simple duration times.

PREDECESSOR \ SUCCESSOR	START	A	E	G	H	B	I	D	F	C	FINISH
START	0	0				0				0	
B	0	20					20				
C	0										30
H	35									16	
A	0							16			
D	20		15	15							
F	16						15				
G	35						3				
E	20						10				
I	38										12
FINISH	51										

Step 5

A forward pass is now carried out to find the earliest starting time of the various activities. This is done as follows—

(*a*) if the project can start NOW insert o in the START—START cell, otherwise insert the start date, calculated in elapsed time, from NOW as o.

(*b*) Search the START row for any "boxed" cells, and add the numbers within these cells to the START—START number. Enter the results in the START column beside the activity under which the cell had appeared.

Step 6

Search across the rows which have just been labelled for boxed cells. For example in Row B, which has just been labelled o, there are two boxed cells, one in the E column, one in the D column. Since the rest of the columns under the headings E and D are vacant, the number in each of the boxes is added to the number in the START column, and the totals entered in the row titled with the same description as the column heading. Thus, the B—E cell contains a value 20 and as there are no other entries in the E column this is added to o (o + 20 = 20) and this total placed in the START column against E. A similar calculation is carried out for the B—D cell and an entry made against D.

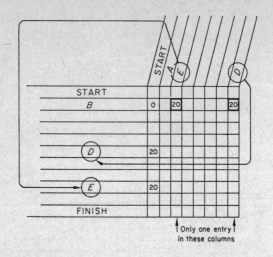

Only one entry in these columns

		SUCCESSOR										
				16	10	3	6	20	12	15	15	30	
PREDECESSOR			START	A	E	G	H	B	I	D	F	C	FINISH
	START		0				0				0		
	B	20		20				20					
	C	30										30	
	H	16										16	
	A	16							16				
	D	15			15	15							
	F	15								15			
	G	3								3			
	E	10								10			
	I	12										12	
	FINISH												

Step 7

Once a boxed cell has been located it may be that the column containing that box holds other boxes, for example, the C—Finish box. Two other boxes are in that column, the H—Finish box and the I—Finish box. Each box is added to the value opposite it in the START column, but if no value appears in the START column, then this process is delayed until an entry does appear. When as many START column cells are filled as possible, the whole process is repeated until all START cells are filled. Where there are several entries in a column, each is added to the START cell and the largest is placed in the row labelled with the column heading.

The figures now appearing in the START column give the earliest times for the activities immediately adjacent on the left, the figure against FINISH giving the total project time.

	START	A	E	G	H	B	I	D	F	C	FINISH
START	0	0			0					0	
B	0		20			20					
C	0										30
H											16
A	0							16			
D	20			15	15						
F	16					15					
G						3					
E	20					10					
I											12
FINISH											

Columns labelled (SUCCESSOR, diagonal headings): START, A, E, G, H, B, I, D, F, C, FINISH. Rows labelled (PREDECESSOR): START, B, C, H, A, D, F, G, E, I, FINISH.

Step 8

A backward pass is now carried out to find the latest starting times of the various activities. This is done as follows—

(*a*) If the project time found above is acceptable, insert it in the FINISH—FINISH cell, otherwise insert the acceptable project time.

(*b*) Search the Finish column for any "boxed" cells and subtract their values from the FINISH—FINISH number. Enter these results in the Finish row under the column heading which is the same as the "boxed" row description. Thus, there is a boxed cell *I*—FINISH of value 12. Subtract this from the FINISH—FINISH value (51 − 12 = 39).

Enter this result under the I column. Similarly, enter $51 - 16 = 35$ under the H column and $51 - 30 = 21$ under the C column.

PREDECESSOR \ SUCCESSOR	START	A	E	G	H	B	I	D	F	C	FINISH
START	0	0				0				0	
B	0		20					20			
C	0										30
H	35										16
A	0								16		
D	20			15	15						
F	16						15				
G	35						3				
E	20						10				
I	38										12
FINISH	51				35		39			21	51

(c) Using the new values in the FINISH row repeat the subtracting process. When a row contains more than one "box" then all values must be subtracted from the FINISH row values, and the smallest entered under the appropriate column heading. Thus, referring to the last diagram the I column has three boxes (15, 3, 10) and none of these have boxes in the same rows, hence three entries (a) can be made—

(i) from row F $39 - 15 = 24$ entered under F

(ii) from row G $39 - 3 = 36$ entered under G

(iii) from row E $39 - 10 = 29$ entered under E

There are now two boxes in row D each of which has a FINISH entry. Hence the entry to be placed under D is the smaller of—

$$35 - 15 = 20 \quad \text{or} \quad 36 - 15 = 21$$

and the value 20 is placed in the FINISH—D cell (entry b).

Similarly, the B row has two boxes and the value entered under B is the smaller of—

$$29 - 20 = 9 \quad \text{or} \quad 20 - 20 = 0$$

that is, 0 is entered at FINISH—B (entry c). The final value (under A) is the difference between 24 and $16 = 8$ and this is entered at FINISH—A (entry d).

The figures appearing in the FINISH row gives the latest starting times for the activities at the column headings.

PREDECESSOR \ SUCCESSOR	START	A	E	G	H	B	I	D	F	C	FINISH
START	0	0			0			0			
B	0	20					20				
C	0										30
H	35										16
A	0							16			
D	20		15	15							
F	16						15				
G	35						3				
E	20						10				
I	38										12
FINISH	51	8	29	36	35	0	39	20	24	21	51

d   aiii   aii   c   b   ai

The matrix method, being effectively MoP, can treat with negative constraints. These are placed within the appropriate box, and ignored on the *forward* pass. On the backward pass they are subtracted from the $START$ column and the difference compared with the difference between the positive value and the FINISH cell.

In practice it will be found that the matrix is easy to construct, and the subsequent analysis easier to perform than to describe. It is a configuration ideally suited for the construction of electrical analogues.

Card Networking

An ingenious networking system uses a set of cards as activities. The activity descriptions are written on the cards, along with any other appropriate information (duration, resources . . .). These cards are then laid upon a table and their interdependencies shown by stretched strings or tapes. By this means the whole process of drawing a network is accelerated, and the originator claims participation is increased. This system is completely described in "Speeded Methods of Network Planning," by J. A. Larkin (*The Production Engineer*, February, 1968).

Other Activity-on-Node Systems

Neither nomenclature, nor practice has yet "hardened" in A-on-N work, and any attempted definitions must be assumed to be tentative.

1. PRECEDENCE DIAGRAMS

The term "Precedence diagram" appears to have been first used by IBM in the 1440 Project Control system. As displayed there it seems to be little or no different from the MoP. A later program "System/360 Project Control System" appears to have elaborated the original concepts in that relationships between activities are shown by particular ways of drawing the arcs and nodes. Three special relationships are identified—

(*a*) The Start of *B* depends on the start of *A* (a start-to-start relationship). This is coded *S* in the "Precedence Relationship" element of the input data, and is drawn—

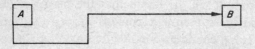

(*b*) The finish of *B* depends on the finishing of *A* (an end-to-end relationship). This is coded *F* and drawn—

(*c*) The start of *B* depends on the finishing of *A* (a start-to-end relationship). This is coded—and drawn—

No constraints (either positive or negative) are mentioned in this program description.

2. THE CIRCLE NOTATION

In this the activity is enclosed within the node circle, along with the activity duration, and these duration times are used in carrying out the forward and backward passes (unlike MoP as discussed above where they are effectively part of the activity description). The subscripts placed on the depending arrows indicate the degree of overlap, thus—

Situation
B can immediately follow *A*

B can start 4 time units before the completion of *A*

B must not start until at least 8 time units have expired after the completion of *A*

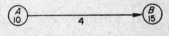

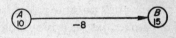

In the forward pass the restraint (or overlap) time is subtracted from the duration of the predecessor in calculating the earlier starting time of a successor, and in the backward pass the restraint time is added to the latest starting time of the successor when calculating the latest starting time of the predecessor.

CHAPTER FOURTEEN

LINE OF BALANCE

HISTORICALLY, Line of Balance (LoB) was developed before CPA, and the two systems are often considered to be separate but related techniques. However, if the original time-scaled stage-time diagram is abandoned, then LoB can be seen to be a quite conventional CPA system applied to a "single-batch" situation.

Where LoB Can Be Used

Just as CPA is used to schedule and control a single project, LoB can be used to schedule and control a single batch. The following requirements need to be satisfied—

(*a*) there must be identifiable stages in production at which managerial control can be exerted;

(*b*) the manufacturing times between these stages must be known;

(*c*) a delivery schedule must be available;

(*d*) resources can be varied as required.

Whilst it is possible to use LoB to control a number of separate batches, just as it is possible to use CPA to control a number of separate projects, the computational difficulties become great. It is therefore usual to emply LoB in "single batch" situations where the batch concerned is of some considerable importance to the organization. An estate of houses, a batch of guided weapons, a batch of computers, are likely to be the type of work appropriate to LoB control.

LoB in Use

The LoB technique will be illustrated by reference to the following hypothetical example—

Product Z is assembled from five components, A, B, C, D and E. A is purchased outright and B is made, tested and then joined with A to make Sub-Assembly 1 (S/A1). C is also made and tested, and then assembled with S/A 1 to give sub-assembly 2 (S/A 2). The material for D has to be purchased, and it can then be made up and tested, and then joined with S/A 2 to give sub-assembly 3 (S/A 3). E is a purchased item which is assembled to S/A 3 at the final assembly stage to give the complete Product Z. This final assembly stage can be considered to include the act of delivering the product to the customer. The delivery schedule is as follows—

First delivery in week ending 1st January.

Week Number	Quantity
1	2
2	4
3	8
4	12
5	10
6	10
7	16
8	18
9	20
10	22
11	24
12	26
13	28
14	24
15	10
16	6
17	4
18	2
19	2
20	2
Total	250

Step 1

Construct a CPA diagram to show the logic and timing of the production. It will usually be found most convenient to start to draw this from the end (in this case "Final Assembly"), and work towards the various opening activities. The network need not be closed at the start—multiple starts are quite permissible and useful here—and nodes need not necessarily be identified, although for the purposes of the present text the nodes are identified here by letters. Duration times indicate the time required for unit production: these times are maintained constant during production by variation of resources. The final chart is now very similar to the "GOZINTO" diagram discussed by—for example—Vaszonyi.

Step 2

Carry out a reverse forward pass from time o at the final event, that is, assign to the final node a time o, and then successively add duration times for each activity in order. This will give the set of figures inscribed against each node, 2 at N, 3 at M, 8 at J and so on.

The result of this reverse forward pass can also be represented on a time-scaled diagram, which is the form in which LoB results are often presented.

Node Times

Whilst the node times represent the latest possible finishing times for the various activities, it is probably more useful to consider these times in relation to the quantities which would pass through the head nodes at any given time. Consider, for example, the activity "Make component B." Any single component B, having been made, will subsequently require three weeks for testing, four weeks to be assembled into S/A1, five weeks to be assembled into S/A2, one week to be assembled into S/A3 and a final two weeks to be incorporated into the final assembly. Therefore, the interval of time in weeks which

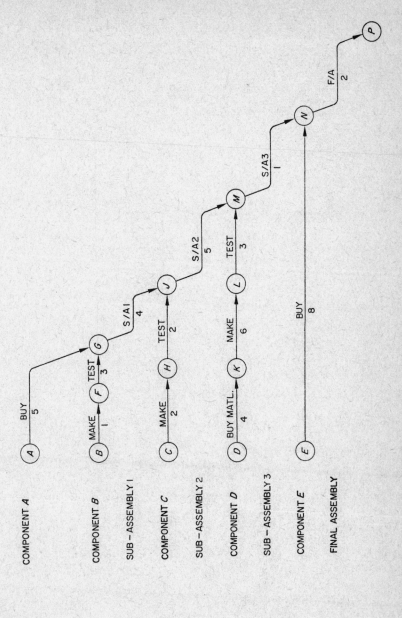

COMPONENT A — (A) — BUY 5 — (G)

COMPONENT B — (B) — MAKE 1 — (F) — TEST 3 — (G)

SUB — ASSEMBLY 1 — S/A1 4 — (J)

COMPONENT C — (C) — MAKE 2 — (H) — TEST 2 — (J)

SUB — ASSEMBLY 2 — S/A2 5 — (M)

COMPONENT D — (D) — BUY MATL. 4 — (K) — MAKE 6 — (L) — TEST 3 — (M)

SUB — ASSEMBLY 3 — S/A3 1 — (N)

COMPONENT E — (E) — BUY 8 — (N)

FINAL ASSEMBLY — F/A 2 — (P)

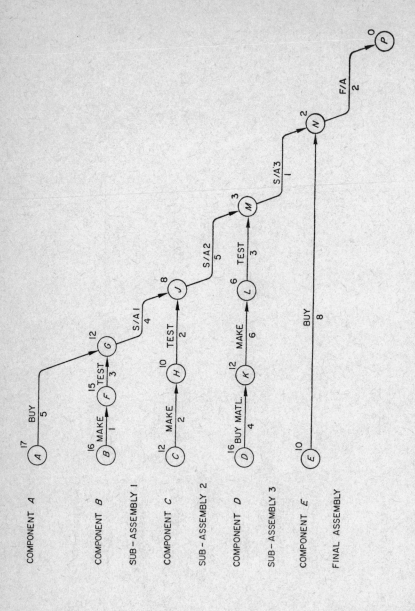

COMPONENT A

COMPONENT B

SUB-ASSEMBLY 1

COMPONENT C

SUB-ASSEMBLY 2

COMPONENT D

SUB-ASSEMBLY 3

COMPONENT E

FINAL ASSEMBLY

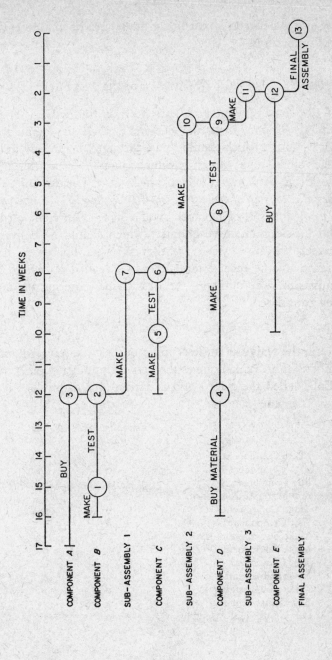

must elapse between a unit being made and its final assembly into Product 2 is—

$$\underset{\text{(test } B)}{3} + \underset{\text{(S/A1)}}{4} + \underset{\text{(S/A2)}}{5} + \underset{\text{(S/A3)}}{1} + \underset{\text{(F/A)}}{2} = 15$$

If the conclusion of Final Assembly is the delivery of the complete Product 2 to the customer, then the cumulative quantity of B's which should "pass through" node F by time t is the cumulative quantity which should "pass through" node P (i.e. be delivered) by a time $t + 15$. For example, two weeks *after the start of delivery of complete "Product Z" to the customer*, the total quantity of B which should have been completed is equal to the cumulative quantity which should be delivered by week $15 + 2 =$ week 17, that is, 244. This node time obtained by the reverse forward pass is called elsewhere the "equivalent week number" for all activities entering the node being considered.

Step 3

Rank the activities in descending order of "equivalent week number." This ranking gives the activity number—sometimes, in LoB, called the stage—and is carried out to produce later a tidy "cascade" chart—

Activity	Equivalent Week Number	Activity Number
Make Component B	15	1
Test Component B	12	2
Buy Component A	12	3
Buy Material Component D	12	4
Make Component C	10	5
Test Component C	8	6
Make Sub-assembly 1	8	7
Make Component D	6	8
Test Component D	3	9
Make Sub-assembly 2	3	10
Make Sub-assembly 3	2	11
Buy Component E	2	12
Carry Out Final Assembly	0	13

TABLE I

Step 4

Prepare a calendar and accumulated delivery quantity table—

Date	Week Number	Quantity	Cumulative Quantity
4th September	—17		
11th September	—16		
18th September	—15		
25th September	—14		
2nd October	—13		
9th October	—12		
16th October	—11		
23rd October	—10		
30th October	—9		
6th November	—8		
13th November	—7		
20th November	—6		
27th November	—5		
4th December	—4		
11th December	—3		
18th December	—2		
25th December	—1		
1st January	1	2	2
8th January	2	4	6
15th January	3	8	14
22nd January	4	12	26
29th January	5	10	36
5th February	6	10	46
12th February	7	16	62
19th February	8	18	80
26th February	9	20	100
5th March	10	22	122
12th March	11	24	146
19th March	12	26	172
26th March	13	28	200
2nd April	14	24	224
9th April	15	10	234
16th April	16	6	240
23rd April	17	4	244
30th April	18	2	246
7th May	19	2	248
14th May	20	2	250

TABLE II

Step 5

From the above two tables deduce the quantity of each activity which should be completed by any particular date. For example—

It is now 22nd January. How many of each component should be completed?
Consider "Make Component *D*."
The time is now week 4.

The quantity through "Make Component *D*" is equal to the quantity which can pass through the final stage in six weeks time that is, in week $4 + 6 = 10$. From Table II this is a total of 122 units.
Similarly, for all the activities—

	Volume of Work Completed is Equivalent to Volume Delivered at Week	Total Units
Make Component *B*	$4 + 15 = 19$	248
Test Component *B*	$4 + 12 = 16$	240
Buy Component *A*	$4 + 12 = 16$	240
Buy Material Component *D*	$4 + 12 = 16$	240
Make Component *C*	$4 + 10 = 14$	224
Test Component *C*	$4 + 8 = 12$	172
Make Sub-assembly 1	$4 + 8 = 12$	172
Make Component *D*	$4 + 6 = 10$	122
Test Component *D*	$4 + 3 = 7$	62
Make Sub-assembly 2	$4 + 3 = 7$	62
Make Sub-assembly 3	$4 + 2 = 6$	46
Buy Component *E*	$4 + 2 = 6$	46
Carry out Final Assembly	$4 + 0 = 4$	26

TABLE III

This can be represented on a chart—the traditional LoB chart.

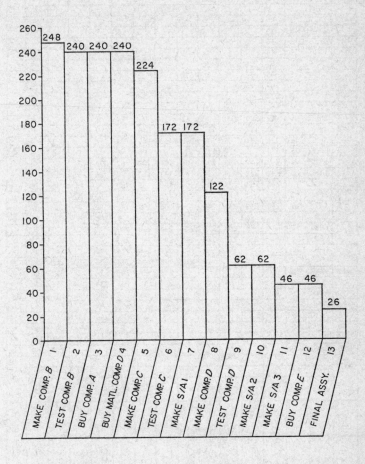

A complete table for the whole "Life" of the batch can be drawn up if desired—see Table IV. The S's in the table indicate the latest dates by which the various chains of activities should START, this date being derived from the equivalent week numbers from the opening activities. The C's in the table show that work must be CONTINUED.

WEEK NUMBER	WEEK STARTING	1 MAKE COMP. B	2 TEST COMP. B	3 BUY COMP. A	4 BUY MATL. D	5 MAKE COMP. C	6 TEST COMP. C	7 MAKE S/A 1	8 MAKE COMP. D	9 TEST COMP. D	10 MAKE S/A 2	11 MAKE S/A 3	12 BUY COMP. E	13 FINAL. ASSY.
−17	SEPT. 4			S										
−16	SEPT. 11	S		C	S									
−15	SEPT. 18	2	S	C	C									
−14	SEPT. 25	6	C	C	C									
−13	OCT. 2	14	C	C	C									
−12	OCT. 9	26	2	2	2	S		S	S					
−11	OCT. 16	36	6	6	6	C		C	C					
−10	OCT. 23	46	14	14	14	2	S	C	C				S	
−9	OCT. 30	62	26	26	26	6	C	C	C				C	
−8	NOV. 6	80	36	36	36	14	2	2	C		S		C	
−7	NOV. 13	100	46	46	46	26	6	6	C		C		C	
−6	NOV. 20	122	62	62	62	36	14	14	2	S	C		C	
−5	NOV. 27	146	80	80	80	46	26	26	6	C	C		C	
−4	DEC. 4	172	100	100	100	62	36	36	14	C	C		C	
−3	DEC. 11	200	122	122	122	80	46	46	26	2	2	S	C	
−2	DEC. 18	224	146	146	146	100	62	62	36	6	6	2	2	S
−1	DEC. 25	234	172	172	172	122	80	80	46	14	14	6	6	C
1	JAN. 1	240	200	200	200	146	100	100	62	26	26	14	14	2
2	JAN. 8	244	224	224	224	172	122	122	80	36	36	26	26	6
3	JAN. 15	246	234	234	234	200	146	146	100	46	46	36	36	14
4	JAN. 22	248	240	240	240	224	172	172	122	62	62	46	46	26
5	JAN. 29	250	244	244	244	234	200	200	146	80	80	62	62	36
6	FEB. 5		246	246	246	240	224	224	172	100	100	80	80	46
7	FEB. 12		248	248	248	244	234	234	200	122	122	100	100	62
8	FEB. 19		250	250	250	246	240	240	224	146	146	122	122	80
9	FEB. 26					248	244	244	234	172	172	146	146	100
10	MAR. 5					250	246	246	240	200	200	172	172	122
11	MAR. 12						248	248	244	224	224	200	200	146
12	MAR. 19						250	250	246	234	234	224	224	172
13	MAR. 26								248	240	240	234	234	200
14	APRIL 2								250	244	244	240	240	224
15	APRIL 9									246	246	244	244	234
16	APRIL 16									248	248	246	246	240
17	APRIL 23									250	250	248	248	244
18	APRIL 30											250	250	246
19	MAY 7													248
20	MAY 14													250

TABLE IV

Step 6

Record the actual progress upon either the LoB chart or the "Life" table. For example, if at 22nd January the achieved and planned results are—

		Achieved	Planned
1	Make Component B	200	248
2	Test Component B	200	240
3	Buy Component A	200	240
4	Buy Material Component D	200	240
5	Make Component C	200	224
6	Test Component C	200	172
7	Make Sub-assembly 1	190	172
8	Make Component D	200	122
9	Test Component D	200	62
10	Make Sub-assembly 2	150	62
11	Make Sub-assembly 3	100	46
12	Buy Component E	90	46
13	Final Assembly	90	26

the LoB chart will be as follows whilst the life table will be as in Table V.

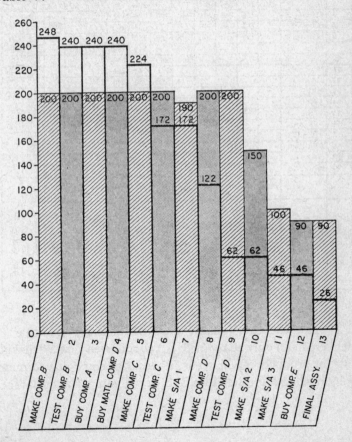

WEEK NUMBER	WEEK STARTING	1 MAKE COMP. B	2 TEST COMP. B	3 BUY COMP. A	4 BUY MATL. D	5 MAKE COMP. C	6 TEST COMP. C	7 MAKE S/A 1	8 MAKE COMP. D	9 TEST COMP. D	10 MAKE S/A 2	11 MAKE S/A 3	12 BUY COMP. E	13 FINAL ASSY.
−17	SEPT. 4	4				8								
−16	SEPT. 11		8			C	8							
−15	SEPT. 18		2			8	C							
−14	SEPT. 25		6			C	C	C						
−13	OCT. 2		14			C	C	C						
−12	OCT. 9	26	2	2	2	8			8	8				
−11	OCT. 16	36	6	6	6	C			C	C				
−10	OCT. 23	46	14	14	14	2	8	C	C				8	
−9	OCT. 30	62	26	26	26	6	C	C	C				C	
−8	NOV. 6	80	36	36	36	14	2	2	C			8	C	
−7	NOV. 13	100	46	46	46	26	6	6	C			C	C	
−6	NOV. 20	122	62	62	62	36	14	14	2	8		C	C	
−5	NOV. 27	146	80	80	80	46	26	26	6	C		C	C	
−4	DEC. 4	172	100	100	100	62	36	36	14	C		C	C	
−3	DEC. 11	200	122	122	122	80	46	46	26	2	2	8	C	
−2	DEC. 18	224	146	146	146	100	62	62	36	6	6	2	2	8
−1	DEC. 25	234	172	172	172	122	80	80	46	14	14	6	6	C
1	JAN. 1	240	200	200	200	146	100	100	62	26	26	14	14	2
2	JAN. 8	244	224	224	224	172	122	122	80	36	36	26	26	8
3	JAN. 15	246	234	234	234	200	146	146	100	46	46	36	36	14
4	JAN. 22	248	240	240	240	224	172	172	122	62	62	46	46	26
5	JAN. 29	250	244	244	244	234	200	200	146	80	80	62	62	36
6	FEB. 5		246	246	246	240	224	224	172	100	100	80	80	46
7	FEB. 12		248	248	248	244	234	234	200	122	122	100	100	62
8	FEB. 19		250	250	250	246	240	240	224	146	146	122	122	80
9	FEB. 26					248	244	244	234	172	172	146	146	100
10	MAR. 5					250	246	246	240	200	200	172	172	122
11	MAR. 12						248	248	244	224	224	200	200	146
12	MAR. 19						250	250	246	234	234	224	224	172
13	MAR. 26								248	240	240	234	234	200
14	APRIL 2								250	244	244	240	240	224
15	APRIL 9									246	246	244	244	234
16	APRIL 16									248	248	246	246	240
17	APRIL 23									250	250	248	248	244
18	APRIL 30											250	250	246
19	MAY 7													248
20	MAY 14													250

TABLE V

Despite the over-fulfilment of the delivery schedule (90 delivered and only 26 required), it can be seen that a "choking-off" of production will occur in weeks to come due to under-fulfilment on some activities, and, equally important, that there is an over-investment in work-in-progress on other activities. It may therefore be possible to transfer resources from the "rich" activities to the "poor" ones whilst preserving the delivery schedule: decisions here can only be taken in the light of local knowledge, and will require reference to both the CPA diagram and the progress results.

Design/Make Projects—Joint CPA, LoB

It is not uncommon to find projects which involve a setting-up stage (design, plan, make jigs and tools) followed by the production of a batch of equipment. Here it is possible to use conventional CPA for all the work up to, and including the making of the first complete equipment, and then to employ LoB to control the subsequent batch production.

CHAPTER FIFTEEN

THE USE OF THE COMPUTER

IT is not appropriate here to discuss the way in which a computer operates. For the purposes of this text it will be assumed that a computer is a device which can carry out arithmetical operations very rapidly, and can file and use information, either arising from the arithmetic or from an outside source as instructed by the program. Any data to be used is read into the computer by the program using an *input* unit and the results are delivered by the *output* unit.

The computer cannot carry out any operation which cannot be carried out manually, given time. Its tremendous advantages all stem from its ability to carry out these calculations extremely rapidly: indeed the limitations on speed are the speeds at which the input and output devices can operate, and continual research is resulting in ever higher speeds. Certainly the present situation is that input and output speeds are more than adequate for all CPA calculations and manipulations.

Preparing a Network for the Computer

The network, once drawn, must be translated into a form with which the computer can deal. As pointed out on page 15, the logic of a network is entirely specified by the event numbers, and to carry out the basic calculations of activity start and finish times and float all that is required is a statement for each activity (including dummies) of the tail number, the head number and the duration time. The first step, therefore, is to prepare a list of these items, which for the sample network so frequently discussed would appear as follows—

Tail Number	Head Number	Duration
1	2	16
1	3	20
1	11	30
2	8	15
3	7	15
3	8	10
7	8	3
7	11	16
8	11	12

From this list ("data list" or "punching list") the input for the computer is prepared. For CPA it is usually more convenient to feed information into the computer by means of punched cards, although other input media can be used. Assuming a card input, a series of cards (a "pack") is prepared, one card for each activity, in which holes are punched in positions which the program can recognize as meaning (for the first activity above): tail number = 1, head number = 2, duration time = 16. (*See* illustration below.) The preparation of the input data is the

(*Courtesy IBM United Kingdom Ltd.*)

This card is punched as follows:
  Columns 3—7: Predecessor event (i.e. tail) number.
  Columns 8—12: Successor event (i.e. head) number.
  Columns 15—42: Description—in this case duration only.

lengthiest and most tedious task associated with using the computer with CPA. Some users avoid preparing the data list by punching the cards directly from the network itself. However, this entails some problems in ensuring that no activities are missed and also requires that the punch operator can read a network.

Instructing the Computer: The Program

Before the prepared data, instructions must be fed into the computer so that calculations can proceed. These are known as a *program*—that is, a set of instructions laying out precisely and in detail the procedures to be followed in transforming input to output—and the preparation of these programs is a skilled task. All computer companies have already prepared substantial libraries of CPA programs which are readily available, and it is extremely rare that a special program needs to be written. The facilities offered by the various programs vary from simple float calculation programs to comprehensive float manipulation programs. Users should decide on the facilities required and select the program which most nearly matches their requirements. Using an unnecessarily complex program can be needlessly expensive.

Facilities Offered in Computer Programs

1. ABILITY TO SORT OUTPUT DATA

Nearly all programs will sort output data, though not all in the same way. The most common output presentations are—

(i) *By order of total float*

This presents the critical activities first, followed by the other activities in order of increasing float, the activities with greatest float appearing last. This has the advantage of directing attention first to those areas where difficulties are most likely to occur, that is, the most critical, and it can thus make monitoring and replanning most economical. (*See* Table I.)

(ii) *By the latest starting date sequence*

Here a diary is effectively constructed in which the earliest entries are those which must be started earliest, and the latest entries those which can be started latest. This enables a ready check to be made to see that activities have been started by the latest acceptable dates. (*See* Table II.)

(iii) *By the latest finishing date sequence*

Here a comparison of actual finishing dates with the latest acceptable finishing dates will indicate any over-running of activities. (*See* Table III.)

Other data presentations are—

(iv) *By the earliest starting date sequence*

(v) *By the earliest finishing date sequence*

But these last two are not generally so useful as the first three presentations.

Within the above sorting, further sorting is possible, for example—

(*a*) by department

(*b*) by resource

(*c*) by responsibility.

These three will require input data additional to the minimum suggested above, and this takes the form of a code (usually numerical) for each sort-type, entered against each activity on the input cards. This second type of sorting makes it possible to obtain, for example, a statement for each department of the latest time by which all the activities carried out in that department should be started (or finished).

2. DUMMY SUPPRESSION

A few computer programs, in order to economize on output, cause dummies to be suppressed in the final presentation. This can be misleading if a dummy is located at the end of a chain, for example—

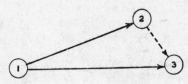

TABLE I: SORTING BY ORDER OF TOTAL FLOAT

Activity	Duration	Start Time		Finish Time		Float		
		Early	Late	Early	Late	Tot.	Free	Ind.
1— 3	20	0	0	20	20	0	0	0
3— 7	15	20	20	35	35	0	0	0
7—11	16	35	35	51	51	0	0	0
7— 8	3	35	36	38	39	1	0	0
8—11	12	38	39	50	51	1	1	0
1— 2	16	0	8	16	24	8	0	0
2— 8	15	16	24	31	39	8	7	0
3— 8	10	20	29	30	39	9	8	8
1—11	30	0	21	30	51	21	21	21

TABLE II: SORTING BY ORDER OF LATEST STARTING DATE

Activity	Duration	Start Time		Finish Time		Float		
		Early	Late	Early	Late	Tot.	Free	Ind.
1— 3	20	0	0	20	20	0	0	0
1— 2	16	0	8	16	24	8	0	0
3— 7	15	20	20	35	35	0	0	0
1—11	30	0	21	30	51	21	21	21
2— 8	15	16	24	31	39	8	7	0
3— 8	10	20	29	30	39	9	8	8
7—11	16	35	35	51	51	0	0	0
7— 8	3	35	36	38	39	1	0	0
8—11	12	38	39	50	51	1	1	0

TABLE III: SORTING BY ORDER OF LATEST FINISHING DATE

Activity	Duration	Start Time		Finish Time		Float		
		Early	Late	Early	Late	Tot.	Free	Ind.
1— 3	20	0	0	20	20	0	0	0
1— 2	16	0	8	16	24	8	0	0
3— 7	15	20	20	35	35	0	0	0
7— 8	3	35	36	38	39	1	0	0
2— 8	15	16	24	31	39	8	7	0
3— 8	10	20	20	30	39	9	8	8
7—11	16	35	35	51	51	0	0	0
8—11	12	38	39	50	51	1	1	0
1—11	30	0	21	30	51	21	21	21

Free float appears at the *end* of chains, and dummy suppression could thus present a situation where no free float was shown. If a dummy suppression program is used, dummies should be located as—

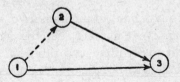

3. ERROR DETECTION

Programs can only detect errors resulting from a logical inconsistency in the input data. Errors arising from illogic in the original network—"loops" and "dangles"—are usually readily detected and signalled. Some programs even print out the event numbers associated with these errors thus simplifying their identification and correction.

Much more difficult to detect is the error which arises from a transcription fault which does not offend the basic logic of the network—for example a time duration of "5" written as "3," or an arrow which, entering at a wrong event, does not produce a loop. To detect these, special input validation routines can sometimes be devised—event numbers can be coded to have a logic of their own (e.g. all are multiples of the same number) and if this logic is not obeyed during transcription an error is signalled. Such validation techniques, though feasible, are usually clumsy or unwieldy, and the author has not discovered a published program which incorporates them. Here, too, much work is being carried out and simplified techniques will doubtless soon be readily available.

4. RANDOM EVENT NUMBERING

One of the most irritating tasks in constructing large networks is the requirement to keep head numbers larger than tail numbers. Most computer programs now accept events numbered in any sequence ("random numbering"), and carry out a sorting of activities internally. This is known as "ranking" or "topological sorting" and is carried out before the forward and backward passes are made. This increases the time on the computer (and hence the cost), but the increase is so small as to be negligible.

There is generally a limit to the number of digits which can be accepted in an event number, but this is usually so high (6–7 digits) that it rarely presents a problem. Some programs will also accept letters within the event numbers.

5. CALENDAR DATES

Normal calculations produce event times which take as a datum the opening event, so that all times are measured from the start of the project. In practice it is often more convenient to refer to calendar dates, and some programs can produce printouts with calendar-date times. When examining a program for this feature it is well to check—

(*a*) the length of the working week used.

(*b*) the ability to allow for holidays.

(*c*) the type of print-out—some programs print out dates in the American fashion (month, day, year) and some in the British fashion (day, month, year).

6. SCHEDULED DATES

Some programs will accept a particular date for the final event, and some a particular date for the final *and* intermediate events. These particular dates are known as *scheduled* dates (*see* page 71) and their use can produce positive, negative and zero floats. When intermediate scheduled dates are used as well as final dates, two or more sets of floats may be produced, and discontinuous critical paths may appear.

7. MULTIPLE STARTS AND FINISHES

Many programs will accept multiple starts and finishes, although some time relationships between the various events may need to be specified. The use of this facility may well simplify the construction of the network but it may also engender a slip-shod approach, and thus care should be taken in specifying more than one start or finish. It is always worth while ensuring that the start and finish of the project being planned are firmly and unambiguously defined.

8. Interfacing

Where activities are carried out in different responsibility areas, some programs permit separate networks to be prepared for each area and these are then combined for the purposes of calculation, provided the common events/activities (the interfaces) are specified. Programs capable of interfacing are usually also capable of printing out separately the results of the separate networks. Thus, in a large project, sub-contractors can produce their own networks, and all the processing can be done simultaneously, separate print-outs subsequently being provided for each sub-contractor. The use of this facility can cause very large networks to appear (for example, 68 sub-contractors, each with a 1,000 activity network, gives a total effective network of 68,000 activities) and great caution should be exercised to ensure that the production of these networks does not effectively produce a shift in operating responsibility which is managerially undesirable.

9. Multiproject Analysis

Just as subnetworks for a single project can be interfaced, so the networks for independent jobs through a single work station can be amalgamated. The problem is made more difficult since, if the jobs *are* independent of each other, then there are no natural interface events, and these have to be provided. If the resultant amalgam is used for resource manipulation (see below) then it will be necessary to rank the constituent jobs in order of priority. The provision of sensible interfacings, and the ranking of jobs are usually extremely difficult tasks to carry out in practice. As with interfacing, large networks can apparently be produced, and most programs provide for the provision of separate print-outs for each job.

10. RESOURCE AGGREGATION

A statement of the total resources required at any one time, usually calculated when all activities start as early as possible, is provided by some programs, assuming of course that the resource requirements are coded into the input. There is usually a limitation on the number of resources which can be handled in this way, up to 10 being quite common although some programs handle considerably more.

11. RESOURCE LEVELLING

If the resource capacity is known some programs can "level" the resources by—

(1) Ranking activities in order of criticality.

(2) Starting all activities as early as possible.

(3) Summing the resource requirements starting with the most critical activity, and when the resource ceiling is reached, shifting all remaining associated activities within their float until a time can be found when the sum of the resources does not exceed the available capacity. If such a time is not found, then an excess is printed, and the analyst must then re-examine the network to see what modifications may be necessary to reduce the load to the available capacity. This re-examination may well result in a change of the logic of the network.

This type of procedure will "clip" the tops off loads, but will not necessarily reduce the required capacity to a minimum, although several passes, each with a diminished ceiling, can go some way towards this. If it is possible to schedule activities ". . . within the limits of their total floats, such that fluctuations in resource requirements are minimized" (BS 4335:1968, 405), then *resource smoothing* is said to have been carried out. The demonstration that a levelling program has, in fact, produced *minimum* fluctuations, rather than *diminished* fluctuations, is a task of some considerable difficulty.

12. RESOURCE ALLOCATIONS

The sharing of resources between activities in a way other than described above is not at present generally possible. It may be worth quoting *News from C.E.I.R. Limited*, Number 2 (July 1964), where, in a discussion on this subject it is stated that computers have not so far been used to any great extent to perform resource allocations. " . . . The logic of what has to be done is complex from a mathematical point of view, and it is likely to be some years before procedures are evolved which allow us to obtain formally-acceptable solutions. There are problems here, both in the development of mathematics and in methods of computation."

The resource allocation problem is difficult for a single network: for a number of independent networks sharing the same resources the problem is intractable. Thus, as at present developed, CPA is not useful for batch production unless some arbitrary decisions on resource distribution between batches are made. Much work is being carried out on this topic at the moment, and some useful special solutions are available. All these require that special sets of decision rules should be generated to meet the particular case being considered, and these rules need to be expressed numerically.

13. VARIABLE COSTS

Programs which can accept "normal" and "accelerated" costs and calculate which activity(ies) can be reduced most economically have been written, and are generally known as minimum cost expediting programs. The author of these notes has never yet found a user who has employed this facility, probably due to the difficulty in deriving the basic data.

14. Cost Totalling

The totalling of costs expended upon completed activities can be carried out by some programs. These totals can be compared with estimated costs thus providing a cost control system. These are often known as PERT-cost programs.

15. Bar Charts

Some programs will produce bar charts, usually over limited time periods—for example, one month. These charts are initially drawn in the "earliest start" position.

16. Activity Numbering/Coding

In order to cut down the use of dummies, some programs will accept activity numbers so that two activities with the same head and tail event can be numbered differently and thus differentiated. Similarly, activities can be numbered/coded in order that print-outs do not carry descriptions of the various activities. In this way secrecy can be preserved.

17. Three-time Estimates

Three-time estimates can be used for two purposes—

(*a*) to obtain, by averaging, a single time estimate;

(*b*) to calculate the probabilities associated with particular events and scheduling times. A number of programs can accept three-time estimates for the first purpose. Fewer programs can carry out the probabilistic calculations.

Capacity of a Computer

The capacity of programs varies enormously from a few hundred activities up to very many thousands, the largest presently available being 70,000 activities.

When to Use a Computer

For a single calculation of float, it is probable that manual calculation is as fast as total processing time with computer (that is, translation of network + preparation of input material + calculation + print-out) for up to 400 activities. Between 400 and 600 activities, the advantages seem small either way, and above 600 activities the computer shows progressively more and more saving. However, the criterion of number of activities is not particularly useful: it is the number of times that calculations need to be carried out which is important. Thus, a small network which needs to be calculated frequently (either for updating, resource allocation, simulation, policy testing or various sortings) would probably justify the use of a computer better than a large network which is calculated only once.

The cost of using a computer is small, whether a domestic computer is used or time is bought from a computer bureau. Moreover, as salaries increase, and computers increase in performance whilst reducing in price, the difference between manual and computer calculations costs will rapidly disappear. A more important consideration is the tedium of producing the input data: this requires very great care to avoid transcription errors. When calculating manually there is an element of self-checking as the analyst will have considerable familiarity with the project in the network, and will recognize any gross errors which arise.

Facilities Offered in Computer Programs

1. Output sorting.
2. Dummy suppression.
3. Error detection.
4. Random event numbering.
5. Calendar dating.
6. Scheduled dating.
7. Multiple starts and finishes.

8. Interfacing.
9. Multiproject analysis.
10. Resource aggregation.
11. Resource "levelling".
12. Minimum cost expediting.
13. Cost totalling.
14. Bar charts.
15. Activity numbering/coding.
16. Three-time estimate manipulation.

Notes. (i) Not all the above facilities are offered in any one program.

(ii) New programs are continually being written extending the facilities offered.

Activity-on-Node Programs

Whilst the programming problems for Activity-on-Node systems are very similar to those for conventional CPA, the effort expended on them has been much less, with the result that far fewer programs are, at present, commercially available. Some programs can be obtained from the Continent, and most of the above facilities can be offered.

It is probable that the very high capital costs involved in program writing will cause conventional Activity-on-Arrow CPA to be used rather than AoN, since the total investment on CPA programs is extremely high.

Line-of-Balance Programs

Your author is unaware of any commercially available LoB programs, although some individual workers have written and used "domestic" programs.

ANALYSES AND GANTT REPRESENTATIONS OF NETWORKS

(*See* pages 72 and 73)

ANALYSIS OF NETWORK *A*

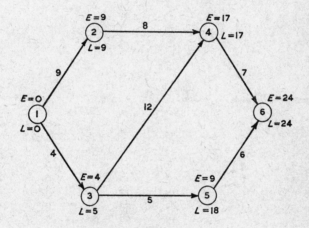

Activity	Duration	Start		Finish		Float		
		Early	Late	Early	Late	T	F	I
1—2	9	0	0	9	9	0	0	0
1—3	4	0	1	4	5	1	0	0
2—4	8	9	9	17	17	0	0	0
3—4	12	4	5	16	17	1	1	0
3—5	5	4	13	9	18	9	0	0
4—6	7	17	17	24	24	0	0	0
5—6	6	9	18	15	24	9	9	0

Critical path: 1—2—4—6

Analysis of Network B

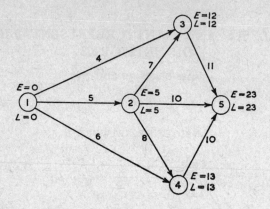

Activity	Duration	Start		Finish		Float		
		Early	Late	Early	Late	T	F	l
1—2	5	0	0	5	5	0	0	0
1—3	4	0	8	4	12	8	8	8
1—4	6	0	7	6	13	7	7	7
2—3	7	5	5	12	12	0	0	0
2—4	8	5	5	13	13	0	0	0
2—5	10	5	13	15	23	8	8	8
3—5	11	12	12	23	23	0	0	0
4—5	10	13	13	23	23	0	0	0

Notes: (1) There are two critical paths—

$$1—2—3—5$$
$$1—2—4—5$$

(2) In this case it is not possible to say that the critical path lies between events whose E and L numbers are the same. *The only safe test is the absence of float.*

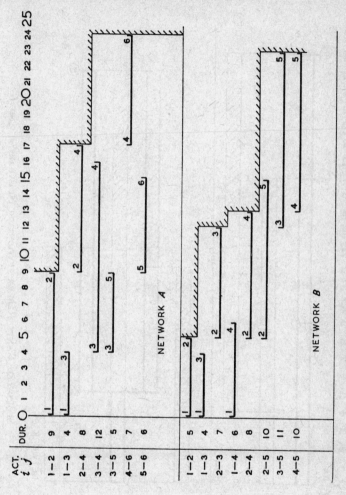

GANTT REPRESENTATIONS OF NETWORKS *A* AND *B*

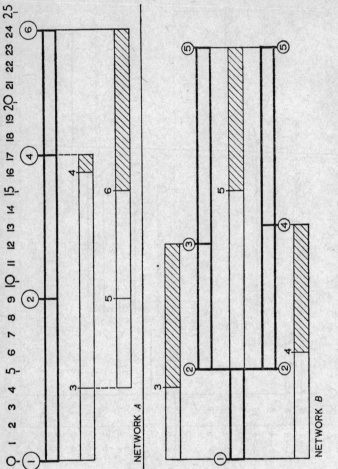

SEQUENCED GANTT REPRESENTATIONS OF NETWORKS A AND B

ANALYSIS OF NETWORK C

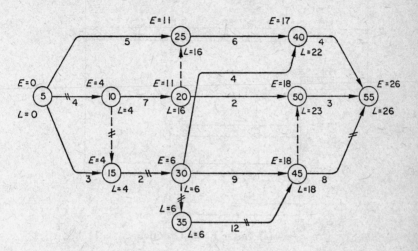

Activity	Duration	Start		Float		Float		
		Early	Late	Early	Late	T	F	I
5—10	4	0	0	4	4	0	0	0
5—15	3	0	1	3	4	1	1	1
5—25	5	0	11	5	16	11	6	6
10—15	0	4	4	4	4	0	0	0
10—20	7	4	9	11	16	5	0	0
15—30	2	4	4	6	6	0	0	0
20—25	0	11	16	11	16	5	0	0
20—50	2	11	21	13	23	10	5	0
25—40	6	11	16	17	22	5	0	0
30—35	0	6	6	6	6	0	0	0
30—40	4	6	18	10	22	12	7	7
30—45	9	6	9	15	18	3	3	3
35—45	12	6	6	18	18	0	0	0
40—55	4	17	22	21	26	5	5	0
45—50	0	18	23	18	23	5	0	0
45—55	8	18	18	26	26	0	0	0
50—55	3	18	23	21	26	5	5	0

Notes: (1) Critical Path 5—10—15—30—35—45—55.
(2) Dummies may possess float, *see* 20—25, 45—50.

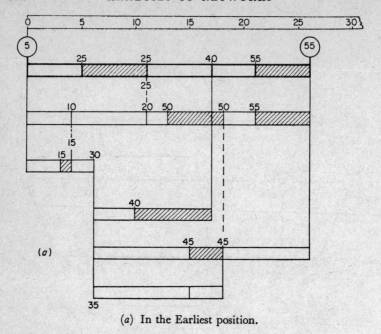

(a) In the Earliest position.

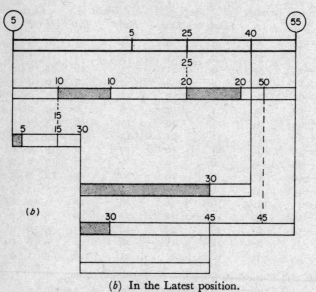

(b) In the Latest position.

SELECTED READING

Despite the comparatively short history of the various network techniques, the volume of written material which has been produced is enormous. Much of this published material has appeared in the form of articles and papers in the various technical journals, and for these reference should be made to the three bibliographies quoted below. At the same time a great deal of useful information has appeared in the publications of computer companies and management consultants, and these, though valuable, should be approached with care, since many of them are often written with a purpose other than that of instructing the reader.

GENERAL

Three fundamental works which should be in the libraries of all serious students are—

1. Special Projects Office.
PERT—Program Evaluation Research Task, Summary Report, Phase I.
2. Special Projects Office.
PERT—Program Evaluation Research Task, Summary Report, Phase II.
Both the above are published by The Department of the Navy, Washington 25, D.C., U.S.A.
3. Office of the Secretary of Defence and National Aeronautics & Space Administration.
D.O.D. and N.A.S.A. Guide, PERT COST Systems Design. (Obtainable from the U.S. Printing Office, Washington 25, D.C., U.S.A.)

Other texts dealing principally with network analysis are—
4. Stilian, Gabriel N., and others.
PERT: A New Management Planning and Control Technique. (The American Management Association, New York, U.S.A.)
This work is particularly useful in that it brings together the experiences of a large number of people who have had practical experience in using networks. There is some overlapping between the various contributions which readers may find irritating, but the total effect is valuable. The book includes a substantial bibliography.
5. Stires, David M., and Murphy, Maurice M.
Modern Management Methods, PERT and CPM (Industrial Education International Ltd., 66 Chandos Place, London, W.C.2)
This was originally written as a teaching text, and is very lavishly illustrated with visual aids. It is clear and comprehensive, and deals at length with the role of the computer. It includes a number of computer print-outs.

6. Fondahl, J. W.

A Non-computer Approach to Critical Path Method for the Constructional Industry. (The Department of Civil Engineering, Stanford University, California, U.S.A.)

As the title suggests, this work deals with the manual manipulation of networks. It is clear and concise, and its usefulness is far greater than the limitations suggested by the words "for the Constructional Industry." It contains a discussion of an "activity-on-node system."

7. Battersby, A.

Network Analysis for Planning and Scheduling (Macmillan & Co. Ltd., London)

As with all Mr. Battersby's works, this is extremely lucid and readable. It covers much the same ground as this book but with some treatment of the more sophisticated techniques. It also contains a number of examples for the reader to work out.

8. Woodgate, H. S.

Planning by Network (Business Publications Ltd.)

A comprehensive work which derives greatly from its author's experience within one of the largest of the computer companies. The role of the computer is fully discussed and illustrated.

9. Miller, Robert W.

Schedule, Cost and Profit Control with PERT (McGraw-Hill Book Co. Inc.)

A useful book which discusses line of balance as well as more general network techniques. The "three-time estimate" technique is treated fully.

10. Moder, J. J. and Phillips, C. R.

Project Management with CPM and PERT (Reinhold Publishing Corporation, New York, and Chapman & Hall Ltd., London)

A useful book which covers not only the introductory material but also some of the newer methods of allocating resources. The availability and use of some 50 different computer programs is discussed.

11. Lockyer, K. G.

Critical Path Analysis: Problems and Solutions (Sir Isaac Pitman & Sons Ltd.)

A series of graded problems in CPA ranging from the very slight to the highly complex. Answers in the form of specimen networks, analyses, bar charts and various manipulations are given.

12. Shaffer, L. R., Ritter, J. B. and Meyer, W. L.

The Critical Path Method (McGraw-Hill Book Co. Inc.)

This describes both conventional arrow diagrams and the circle diagram technique whereby the use of dummies can be avoided. The use of computers is discussed, and there is a substantial treatment of the "cost-slope" technique.

13. O'Brien, James J.
CPM in Construction Management (McGraw-Hill Book Co. Inc.)
As indicated by the title, all the illustrations and examples are derived from the Construction industry. The approach is extremely practical and demonstrates the author's wide experience.

14. Archibald, R. and Villoria, R.
Network based Management Systems (John Wiley & Sons Inc.)
A wider-ranging text which includes discussions of decision trees, line of balance and activity-on-node system. It includes a condensation of the U.S. Government PERT/COST report formats.

15. Mantino, R. L.
Project Management and Control Vol. 1 : Finding the Critical Path. Vol. 2: Applied Operational Planning. Vol. 3: Allocating and Scheduling Resources (American Management Association, New York)
Three excellent books by one of the pioneers of this subject.

16. Horowitz, J.
Critical Path Scheduling: Management Control through CPM and PERT. (Ronald Press Co., New York).
The text includes a number of problems for the reader to solve. Many of the illustrations are derived from the construction industry, but workers in other fields will find it useful.

17. McLaren, K. G. and Bueonel, E. L.
Network Analysis in Project Management (Cassell & Co. Ltd.)
A summary of the authors' wide experience in teaching and applying the techniques within Unilever. A very large format enables several practical networks to be reproduced.

18. Brennan, Jas (ed.)
Applications of Critical Path Techniques (English Universities Press Ltd.)
A collection of 23 papers with ensuing discussion by users of Critical Path Techniques. These were delivered at a conference held under the aegis of the NATO Scientific Affairs Committee in Brussels, July/August, 1967.

19. Lock, D.
Project Management (Gosver Press Ltd.)
An attempt to consider all aspects of Project Management including CPA. Again, many illustrations are from the construction industry.

20. Mulvancy, J.
Analysis Bar Charting (Iliffe Books Ltd.)
A simplified networking technique based upon an Activity-on-Node system. It includes a discussion of Line-of-Balance.

BIBLIOGRAPHIES

There are two very useful and comprehensive bibliographies which those wishing to study the origins and development of the subject should consult—

1. Sobczac, Thomas V.
Network Planning—a Bibliography. The Journal of Industrial Engineering, Vol. 13, No. 6, November-December, 1962.

(The American Institute of Industrial Engineers Inc., 345 East Forty-Seventh Street, New York 17, New York, U.S.A.)

2. *A Bibliography of CPM and PERT.*
Industrielle Organisation (February, 1963), Postfach, Zurich, Switzerland.

3. *A Bibliography of Critical Path Methods* (1966).
Hampshire Technical Research Industrial Commerical Service, Central Library Civic Centre, Southampton.

READING IN ASSOCIATED AREAS

As a managerial tool, CPA is often used in conjunction with other techniques, and reading in these other subjects is essential. Some books which the author has found useful are listed below; they have been classified according to the chapter of the present work to which they relate.

Chapter 8. *Reducing the Project Time*
The material covered by this section is normally dealt with under the title "Work Study," or "Time and Motion Study." Two excellent readable general works are—
1. International Labour Office.
   Introduction to Work Study.
2. Currie, R. M.
   Work Study (Sir Isaac Pitman and Sons Ltd., for the British Institute of Management)
   A much more comprehensive and detailed work is—
3. Barnes, R. M.
   Motion and Time Study (John Wiley & Sons Inc.)

Chapter 9. *The Arrow Diagram and the Gantt Chart*
Here, of course, the definitive work is—
Clark, Wallace.
The Gantt Chart (Sir Isaac Pitman and Sons Ltd.)
which is written with tremendous enthusiasm and exuberance by a close personal friend of Henry Gantt.
An excellent discussion of the translation of the arrow-diagram into a Gantt chart is given in—
Burgess & Killebrew, Journal of Industrial Engineering, Volume XIII, Number 2, March–April, 1962.

Chapter 10. *Loading I—The General Problem*
This area contains problems which at present are not capable of general solution. A number of special solutions for particular cases can be found, but their integration with CPA may not always be obvious or simple. The author (K.G.L.) has found—

1. Magee, J. F.
   Production Planning & Inventory Control (McGraw-Hill Book Co. Inc.)
to be readable, useful and written from an essentially practical viewpoint.
Also on this subject are—

2. Eilon, Samuel.
Elements of Production Planning and Control (The Macmillan Company, New York)
3. Vazsonyi, A.
Scientific Planning in Business and Industry (John Wiley & Sons Inc.)
4. Bowman, E. H. and Fetter, R. B.
Analysis for Production Management (Richard Irwin Inc.)
5. Elmagrahby, S. E.
The Design of Production System (Reinhold Publishing Corporation)
6. Conway, R., Maxwell. W. and Miller, L.
Theory of Scheduling (Addison-Wesley Publishing Company)
These texts between them probably contain all that has yet been made public on the subject. All require mathematics of a very high standard and all need considerable work on the part of the reader.

The work of the American mathematician, Richard Bellman, is undoubtedly destined to play an increasingly important part in this field. Of his works—
7. Bellman, Richard.
Adaptive Control Processes (Oxford University Press)
will possibly be the most influential. It is lucidly written and extremely stimulating but, once again, requires mathematical understanding of a high order.

Chapter 11. *Loading II—A Manual Method*
The present chapter draws heavily upon the work of D. Martino cited in reference 15, Chapter 1.

Chapter 12. *Control and Critical Path Analysis*
This subject is often hidden under the name "Cybernetics." The work of Norbert Wiener stands out in this field, and reference can usefully be made to any of his works on this subject. Suggested readings are—
1. Wiener, Norbert.
Cybernetics (M.I.T. Press and John Wiley & Sons Inc.)
2. Beer, Stafford.
Cybernetics and Management (English Universities Press)
3. Straffurth, C.
Project Cost Control using Networks (Published jointly by The Operational Research Society, 62 Cannon Street, London E.C.4, and The Institute of Cost and Works Accountants, 63 Portland Place, London W.1. N 4AB.
The collective experience of a number of users of CPA.

Chapter 14. *Activity-on-Node Networking*
This subject is discussed in several of the general texts mentioned above, and in a number of articles and papers, of which the best known are
1. Roy, B.
Graphes et Ordannancements Operationelle (Revue Française de Recherche Operationelle 6e année, 4e trimestre, pp. 323 *et seq*—1962).

2. Meyer, W. E.

Use of Critical Path Method Simplified by the Introduction of Circle Notation Overlap Feature (Society of Automotive Engineers Journal, March, 1966, New York)

This describes an alternative to M. B. Roy's Method of Potentials.

Chapter 15. *The Use of the Computer*

This is dealt with in many of the references under *General* above, notably numbers 8 and 10.

Chapter 16. *Line-of-Balance.*

1. *Line-of-Balance Technology;* NAVEXOS page 1851 (Washington Department of the Navy, 24th February, 1958)

This is the source book of much of the work on LoB. It is not an easy book to obtain, and can probably only be found in the larger libraries.

2. Lumsden, P.

The Line-of-Balance Method (Pergamon Press Ltd.)

This treats LoB as a logical part of CPA. It deals in particular with the case of quantities delivered at a constant rate, and on this base carries out calculations of resource requirements. It contains summaries of a number of actual cases of the case of LoB.

GLOSSARY OF TERMS

WHEREVER appropriate, the term and its definition as given in BS 4335:1968 are used and appear between inverted commas.

"Activity: an operation or process consuming time and possibly other resources."

Activity Span: the time available for the completion of an activity.

Activity Time: see Duration.

Arrow: the symbol by which an activity is represented.

Arrow Diagram: the statement of the complete task by means of arrows.

Backward Pass: the procedure whereby the latest event times for a network are determined.

Circle: the symbol by which an event is represented.

Cost-slope: the cost incurred in reducing the activity time by unit time. (*Note:* a negative cost-slope indicates that the cost of completing an activity decreases as the activity time decreases.)

Critical Path: that sequence of activities which determines the total time for the task. It is "a path from a start event to an end event, the total duration of which is not less than that of any other path between the same two events."

Critical Path Analysis: a method whereby the policy to be adopted in carrying out a task is represented by a graphical model in which the times necessary for the constituent parts are inserted. The model is analysed, the sequence(s) of times which determine the total project time extracted, and the times available for all constituents calculated. Once the task is in being, comparison of actual times with available times enables control to be exerted on the performance of the task. "The project network analysis technique for determining the minimum project duration."

Dangle (**dangling activity**): an activity whose completion does not give rise either to another activity or to the completion of the whole project.

Dependency Rule: the basic rule governing the drawing of a network. It requires that an activity which depends on another activity is shown to emerge from the head event of the activity upon which it depends, and that only dependent activities are drawn in this way.

"Dummy: A logical link, a constraint which represents no specific operation." In calculations it is most usefully regarded as an activity which absorbs neither resources nor time. (*Note:* dummies are usually represented by broken arrows.)

Dummy Activity: a dummy.

"Duration: the estimated or actual time required to complete an activity."

Duration Time: see duration.

Earliest Event Time: the earliest time by which an event can be achieved without affecting either the total project time or the logic of the network. (*Note:* BS 4335:1968 recommends: "Earliest date of event: The earliest date [or point in time] an event can occur.")

Earliest Finish Time of an Activity: the earliest possible time at which an activity can finish without affecting the total project time or the logic of the network. (*Note:* BS 4335: 1968 recommends "Earliest finish date of an activity: The date [or point in time] before which an activity cannot be finished.")

Earliest Start Time of an Activity: the earliest possible time at which an activity can start without affecting either the total project time or the logic of the network. (*Note:* BS 4335: 1968 recommends: "Earliest start date of activity: The date [or point in time] before which an activity cannot be started.")

"Event: a state in the progress of a project after the completion of all preceding activities but before the start of any succeeding activity."

Event Time: the time by which an event can (or is to be) achieved.

"Float: a time available for an activity or path in addition to its duration (may be negative)." It is essentially a property of activities, and is the difference between the time necessary and the time available for an activity.

Forward Pass: the procedure whereby the earliest event times for a network are determined.

Free Float: the float possessed by an activity which, if used, will not change the float in later activities. (*Note:* BS 4335: 1968 recommends "Earliest date of succeeding event of activity minus earliest finish date of activity.")

Free Float—early: another name for free float.

Free Float—late: the float possessed by an activity when its predecessors and successors are achieved as late as possible.

Head Event: the event at the finish of an activity.

Head Slack: the slack possessed by an event at the head of an activity.

i: the symbol for the event number of a tail event.

"Imposed Date: a date (or point in time) determined by authority or circumstances outside the network, or the fixed point for the time scale of the network."

Independent Float: the float possessed by an activity which, if used, will not change the float in any other activities in the arrow diagram. (*Note:* BS 4335:1968 recommends: "Earliest date of succeeding event minus latest date of preceding event minus activity duration. If negative, the independent float is taken as zero.")

Interface (to): the act of coalescing two or more networks.

"Interface: an event which occurs identically in two or more networks or subnetworks."

Interference Float: a component of float equal to the head slack of an activity.

j: the symbol for the event number of an head event.

Junction: another name for an event.

Latest Event Time: the latest time by which an event can be achieved without affecting either the total project time or the logic of the network. (*Note:* BS 4335:1968 recommends: "Latest date of event: The latest date [or point in time] an event can occur.")

Latest Finish Time of an Activity: the latest possible time by which an activity can finish without affecting either the total project time or the logic of the network. (*Note:* BS 4335:1968 recommends: "Latest finish date of activity: The date [or point in time] after which an activity cannot be finished.")

Latest Start Time of an Activity: the latest possible time by which an activity can start without affecting either the total project time or the logic of the network. (*Note:* BS 4335:1968 recommends: "Latest start date of activity: The date [or point in time] after which an activity cannot be started.")

Loop: a sequence of activities in which a later activity is shown to determine an earlier activity.

Milestone: another name for an event. Sometimes reserved for a major or important event.

Negative Float: the time by which the duration of an activity or chain of activities must be reduced in order to permit a scheduled date to be achieved.

Negative Slack: the time by which the difference between the earliest and latest event times for an event must be increased in order to permit a scheduled date to be achieved.

"Network: a diagram representing the activities and events of a project, their sequence and inter-relationships."

Node: another name for event.

PERT: a name for a network analysis technique formed from the words "PROGRAM EVALUATION and REVIEW TECHNIQUE." Originally requiring the use of three estimates of the duration times, the name is now usually accepted as one of the generic names for network techniques.

"Resource Aggregation: the summation of the requirements of each resource for each time period, calculated according to a common decision rule."

Resource Levelling: the utilization of the available float within a network to ensure that the resources required are appreciably constant.

"Resource Limited Scheduling: the scheduling of activities such that predetermined resource levels are never exceeded, and the project duration minimized."

Resource Optimization: the manipulation of the network to try to ensure that the resources required and available are in balance.

"Resource Smoothing: the scheduling of activities within the limits of their total floats, such that fluctuations in resource requirements are minimized."

Resource Totalling: equivalent to "Resource Aggregation."

Scheduled Date ⎫
Scheduled Time ⎭ equivalent to "Imposed Date."

Secondary Float: when a scheduled date is imposed upon an event which is not a final event, a secondary critical path can appear which is the time-controlling sequence between the start and the scheduled event, or between two scheduled events. Activities not on this critical path but which contribute to the achievement of the event possess float with respect to this secondary critical path, and this is said to be secondary float.

Semi-Critical Path: that path which is next to the critical path when all paths are arranged in order of float.

"Slack: Latest date of event minus earliest date of event. (May be negative). The term slack is used as referring only to an event."

Stage: another name for an event.

Sub-Critical Path: a path which is not critical.

Tail Event: the event at the beginning of an activity.

Tail Slack: the slack possessed by an event at the tail of an activity.

Time Available: another name for "activity span."

"Time Limited Scheduling: The scheduling of activities such that the specified project duration is not exceeded, using resources to a predetermined pattern."

Total Float: the total float possessed by an activity.

"Trading Off": the transferring of resources from one activity to another. This is usually accompanied by changes in duration times, and is carried out to affect the resource distribution.

INDEX